SpringerBriefs in Applied Sciences and Technology

SpringerBriefs present concise summaries of cutting-edge research and practical applications across a wide spectrum of fields. Featuring compact volumes of 50 to 125 pages, the series covers a range of content from professional to academic.

Typical publications can be:

- A timely report of state-of-the art methods
- An introduction to or a manual for the application of mathematical or computer techniques
- A bridge between new research results, as published in journal articles
- A snapshot of a hot or emerging topic
- An in-depth case study
- A presentation of core concepts that students must understand in order to make independent contributions

SpringerBriefs are characterized by fast, global electronic dissemination, standard publishing contracts, standardized manuscript preparation and formatting guidelines, and expedited production schedules.

On the one hand, **SpringerBriefs in Applied Sciences and Technology** are devoted to the publication of fundamentals and applications within the different classical engineering disciplines as well as in interdisciplinary fields that recently emerged between these areas. On the other hand, as the boundary separating fundamental research and applied technology is more and more dissolving, this series is particularly open to trans-disciplinary topics between fundamental science and engineering.

Indexed by EI-Compendex, SCOPUS and Springerlink.

Azman Ismail · Fatin Nur Zulkipli ·
Darulihsan Abdul Hamid · Andreas Öchsner
Editors

Advancements in Industry and Technology

Springer

Editors
Azman Ismail
Malaysian Institute of Marine Engineering Technology
University of Kuala Lumpur
Lumut, Perak, Malaysia

Fatin Nur Zulkipli
College of Computing, Informatics and Mathematics
Universiti Teknologi MARA
Machang, Kelantan, Malaysia

Darulihsan Abdul Hamid
Malaysia Italy Design Institute
Universiti Kuala Lumpur
Kuala Lumpur, Malaysia

Andreas Öchsner
Faculty of Mechanical Engineering
Esslingen University of Applied Sciences
Esslingen am Neckar, Baden-Württemberg, Germany

ISSN 2191-530X ISSN 2191-5318 (electronic)
SpringerBriefs in Applied Sciences and Technology
ISBN 978-3-031-97930-9 ISBN 978-3-031-97931-6 (eBook)
https://doi.org/10.1007/978-3-031-97931-6

This Springer imprint is published by the registered company Springer Nature Switzerland AG
The registered company address is: Gewerbestrasse 11, 6330 Cham, Switzerland

Preface

This book provides a comprehensive exploration of the dynamic landscape of modern advancements across diverse sectors. It is aimed at professionals, researchers, and policymakers as it serves as a valuable resource for understanding and navigating the evolving landscape of industry and technology in the modern era.

Lumut, Malaysia	Azman Ismail
Machang, Malaysia	Fatin Nur Zulkipli
Kuala Lumpur, Malaysia	Darulihsan Abdul Hamid
Esslingen am Neckar, Germany	Andreas Öchsner

Contents

Polytechnic Students' Perceptions Toward Online Examination: A Case Study

Siti Salmah Md. Kassim, Nor Azida Mansor, and Mazlina Md Mustaffa

Abstract The implementation of online learning causes exams to be conducted online as well. Online exams have different impacts on students, either positively or negatively. This study investigates students' perceptions concerning online exams, which have been used since COVID-19. Data was collected using quantitative methods. A descriptive analysis was employed to analyze quantitative data. A survey of 295 polytechnic students was conducted to ascertain their attitudes toward online exams. As a result, suggestions for improvement can be given to the top management in polytechnics to improve the implementation method of the examination in the future.

Keywords Online · Exam · Perception · Student · Polytechnic

1 Introduction

In accordance with technology nowadays, online examination is the practical platform to evaluate student assessment. From 2019 to 2022, COVID-19 strike causes education in every level to shift their platform from face-to-face to virtual or online methods. An online examination system is a web-based examination system in which exams are given online via the internet or intranet utilizing a computer system. The major purpose of this online examination system is to properly evaluate the student using a completely automated method that not only reduces the required time but also produces fast and accurate results [1]. However, the implication of moving from the

S. S. Md. Kassim (✉) · N. A. Mansor · M. M. Mustaffa
Jabatan Teknologi Maklumat dan Komunikasi, Politeknik Ungku Omar, Jalan Raja Musa Mahadi, Ipoh, Perak, Malaysia
e-mail: sitisalmah@puo.edu.my

N. A. Mansor
e-mail: azida@puo.edu.my

M. M. Mustaffa
e-mail: mazlina_mustafa@puo.edu.my

A. Ismail et al. (eds.), *Advancements in Industry and Technology*,
SpringerBriefs in Applied Sciences and Technology,
https://doi.org/10.1007/978-3-031-97931-6_1

conventional method (face-to-face) to the online examination method does influence students either emotionally or physically. From there, it can show how good or bad online examination according to student opinion is and to help educational institutions improve their teaching and learning style. Hence, this study aims to identify the perception of technical and vocational education and training (TVET) students especially Malaysian Polytechnic and Community College toward the new era of assessment using online examination and how this can affect their behavior especially in the process of teaching and learning. Besides that, this study also wants to address the challenges that be faced by students during the online examination. This paper is divided into several parts namely, Literature Review, the Methodology, which explains the method which is implemented to gain results at the end, followed by Results and Discussion. This paper also will conclude with a comprehensive conclusion.

2 Literature Review

This part will discuss the previous studies about implementation online examination, the challenges that educational institutions face in order to execute online examination, and the benefits of implementing online examination.

Online Examination in Malaysian Educational Institutions

Online examination is still infancy in terms of implementation in Malaysian Polytechnics. Curriculum Information Document Online System (CIDOS) is an online platform which is used to conduct online learning, assessment, including continuous assessment and the final exam [2]. By using CIDOS, Malaysia Polytechnic and Community College conducts the online exam for all the courses, including the practical orientation course, which needs to use the simulator as an evaluation tool. CIDOS was implemented in the mid of 2010 where some polytechnic institutions started to venture into the use of LMS in conducting teaching and learning processes and is utilized by most polytechnics in Malaysia in 2011 [3]. Since then, CIDOS evolves with many improvements from the infrastructure of networking, server hosting, distribute learning material, disseminate instructions, collecting assignments, and executing online tasks or quizzes. But now it is moving toward maturity in promising ubiquitous access by shifting to cloud services.

Benefits of Online Examination

Online examination promises many advantages in terms of saving time, effort, and cost. Besides that, in tandem with the rapid development of technology nowadays, online examination supports the needs of education in the 4th industry revolution. As mentioned previously in [4, 5], the integration of IR4.0 in higher education system will help the learning institutions to stay relevant and competitive in the dawning of industry 4.0. Higher learning institutions are urge to revamp their education system and take initiatives to embrace the teaching and learning 4.0. One of the underpinned in teaching and learning 4.0 is the advances use of technology 4.0 such as 3D printing, augmented reality, virtual reality, cloud computing, hologram, biometrics, multi-touch LCD screen, internet of things, artificial intelligence, big data, and QR code and so forth. Human can leverage all benefits to make life easier, hence can progress faster to invent new ideas in the future especially related to teaching and learning processes.

Challenges in Implementation of Online Examination

Every technology comes with its own flaws. Same goes to online examination. Previous studies already identified several challenges in implementing online examination which can be summarize into three main factors; human, organization and technology [3]. Human can be the students, instructor, and administrator of the educational institution. For students, the challenge always come from the integrity matter. Students tend to cheat while doing exam and assessments [6]. The instructor on the other hand struggles with the preparation to adapt with the new teaching style which sometimes cannot meet their expectation. Some might be comfortable to meet face-to-face with students while teaching or having assessment, hence they felt uneasy and doubtful using online methods. From a technology perspective, online examination is always tied to the emerge of IT technology. As well known, IT technology is rapidly changing. In parallel with the changes, all the infrastructure such as hardware, network, and software also needs to be competent with the technology. From here, the organization needs to invest a lot of money to buy the hardware and software.

3 Methodology

This study used descriptive analysis which was employed to analyze the quantitative data. A survey of 295 polytechnic students was conducted to ascertain their attitudes toward online exams. As a result, suggestions for improvement can be given to the top management in polytechnics to improve the implementation method of the examination in the future.

Data Collection

Data is collected from a survey which was distributed among Malaysia Polytechnic students.

Table 1 shows the demography of the respondents. The total respondents that answered the survey questions were 295 students from all over Malaysia, including from peninsular Malaysia and also from west Malaysia (Sarawak).

Survey Question

The survey was conducted through distributing the survey question via email. The questions were developed from adaptation of existing survey questions. Studies such as [6, 7] were referred to adapt the questions about the online examination. However,

Table 1 Demography of the respondents

#	Polytechnic	Number of respondents	Female	Male
1	Besut	25	12	13
2	Balik Pulau	32	16	16
3	Kuching, Sarawak	11	3	8
4	Kuala Terengganu	35	18	17
5	Mersing, Johor	28	17	11
6	Muadzam Shah, Pahang	18	10	8
7	Sultan Abdul Halim Mu'adzam Shah	21	9	12
8	Metro Tasek Gelugor	4	2	2
9	Sultan Azlan Shah, Perak	20	12	8
10	Seberang Perai, Penang	33	20	13
11	Ungku Omar	68	35	33

the questions need to be revised to meet this study objectives which is to investigate students' perceptions concerning online exams in Malaysian Polytechnics' student.

Table 2 depicts the questions that were constructed for this study to gain the idea on the student perception toward online examination. These survey questions are important to address the survey result and this will be discussed in the next section.

Table 2 Items in the survey

#	Item
Q1	Online examination makes me tired
Q2	Duration of online exam is not enough for me to complete it
Q3	I get distracted by things around me (family, pets, devices, internet, phones and computers) while doing online exam
Q4	I feel competent (capable) in doing online exam
Q5	I worry I am going to fail in online exam
Q6	I prefer to answer Multiple Choice Question (MCQ)
Q7	I am able to focus on the online exam
Q8	I enjoy online examination
Q9	I can understand the exam instruction clearly while doing online examination
Q10	I feel comfortable doing online exam without lecturer presence physically

4 Results and Discussion

Essentially, the item was intended to get the perception of respondents in terms of their own feelings toward online learning, which is related to their own situation, surroundings, and also in general about the online examination. As depicted in Table 1, the items are labeled as Q1 till Q10 to make it easier in the analyzing process. Before the results were analyzed, first the reliability of the survey questions is validated.

Before all the collected data is analyzed using SPSS, a cleaning process is done to eliminate any unnecessary data that could affect the result. Table 3 depicts the reliability of the items in the survey questions. Some of the respondents misunderstood the items, and some of them carelessly answered without any concern about the meaning of the items. After cleaning the data, the reliability of the items is shown in Table 4. The value of Cronbach's alpha shows that 0.717 of these survey items are reliable to be answered by the respondents and acceptable.

The results from the data were analyzed using the Statistical Package for the Social Sciences Software (SPSS). As shown in Table 4, the result shows that the mean for each of the items is approaching 3 to 4, which, according to the scale, is a great deal. It shows that most respondents agree that online exams make them tired, afraid of failing, the duration is not long enough, and they get distracted by their surroundings. On the other hand, students also agree that an online exam is interesting; they are able to focus; feel competent to do an online exam; understand the instruction; feel comfortable without the lecturer's presence; and they prefer MCQ type questions while doing an online exam.

Table 3 Reliability of item

Cronbach's alpha	Cronbach's Alpha based on standardized items	No. of items
0.717	0.722	10

Table 4 Mean and standard deviation of items

Item	Mean	Standard deviation	N
Q1	2.82	0.737	295
Q2	3.24	0.718	295
Q3	2.77	0.742	295
Q4	3.08	0.673	295
Q5	3.28	0.684	295
Q6	3.13	0.688	295
Q7	2.97	0.769	295
Q8	3.36	0.681	295
Q9	3.27	0.696	295
Q10	3.15	0.767	295

From the data analysis, it shows that Malaysia Polytechnic students in general prefer online exams. But, not to say they prefer the online method as compared to face-to-face because some of them still feel distracted by the surroundings and are not used to the online environment. Malaysia Polytechnic students have to use the online exams because of the government's restrictions on movement during COVID-19. But, the model for implementing online exams is different compared to other higher education because the courses offered is also different in terms of curriculum and syllabus. Hence, a thorough study has to investigate more on the model of implementation of an online exam in the future.

5 Conclusion

The solution needs to be raised to top management at Jabatan Pendidikan Politeknik dan Kolej Komuniti (JPPKK) if there is an enforcement on online examination implemented at Malaysia Polytechnic. Other than the issues discovered by this study on the student perception toward online examinations, there is a need for a stable and robust platform to handle the online examinations at Malaysia Polytechnic institutions. This platform must be reliable and secure to avoid any unnecessary circumstances. A networking infrastructure is also the key to successfully implementing online exams.

Acknowledgements Special thanks to all the respondents who play an important role to cooperate the survey responses. The result can help top management to take relevant action to improve on the line assessment platform.

References

1. M.R. Hameed, F.A. Abdullatif, Online examination system. IARJSET **4**(3), 106–110 (2017). https://doi.org/10.17148/IARJSET.2017.4321
2. S. Nurul, M. Mohamad, S. Salam, N. Bakar, *Lecturers' Perceptions and Attitudes Towards the Usage of Online Learning at Polytechnic* (2014)
3. M. Hasnan, M. Mohin, Implementation of LMS-CIDOS in polytechnic English language classroom: issues and challenges. Asian J. Univ. Educ. **17**(4), 527–537 (2021). https://doi.org/10.24191/ajue.v17i4.16253
4. W. Matli, M. Phurutsi, ScienceDirect students' concerns about online remote learning during COVID—students' concerns 19 Pandemic in the 4IR digital. Procedia Comput. Sci. **219**(2022):971–976 (2023). https://doi.org/10.1016/j.procs.2023.01.374
5. M. Tsakeni, Transition to online learning by a teacher education program with limited 4IR affordances. Res. Soc. Sci. Technol. **6**(2), 129–147 (2021). https://doi.org/10.46303/ressat.2021.15
6. N. Iskandar, N. Ganesan, N.S. Eleena, A. Maulana, *Students' Perception Towards the Usage of Online Assessment in University Putra Malaysia Amidst COVID-19 Pandemic* (2021). [Online]. Available: www.questjournals.org

7. H. Ilgaz, G. Afacan Adanır, Providing online exams for online learners: does it really matter for them? Educ. Inf. Technol. (Dordr) **25**(2), 1255–1269 (2020). https://doi.org/10.1007/s10639-019-10020-6

Digitalization of the Marine NAVAIDs Monitoring System in the Straits of Malacca and Malaysian Waters

Muhammad Fadhli Makhtar, Yasmin Mohd. Hasni, Hairizam Albukhari, and Mohd Hisham Rubani

Abstract The paper examines the comprehensive of digitalization system to enhance the marine NAVAIDs system in monitoring of the traffic and aids to navigation (AtoN) in the Straits of Malacca (SOM) and Malaysian waters. The high significance of the SOM as a bustling critical international shipping route ensuring the safety and efficiency of maritime traffic is vital. The digitalization process involves the integration of advanced technologies such as GPS, radar, and remote sensing to improve the accuracy and reliability of data. Utilizing technology from the marine sector, the E-Nav system permits real-time monitoring and remote control of AtoN, enabling quick repair and efficient navigational response. Moreover, the digitalization of AtoN data opens avenues for the development of decision support systems and predictive analytics, empowering stakeholders with enhanced situational awareness and risk management capabilities. Technological advancement optimizes navigational safety and efficiency by facilitating proactive measures, leveraging historical data analysis, and identifying patterns to anticipate potential issues. The findings highlight the substantial benefits of the digitalization efforts, underscoring their potential to revolutionize maritime navigation in the SOM and Malaysian waters and offering valuable sights for future developments in the domain of maritime traffic management.

Keywords Aids to navigation · Digitalization · Monitoring · Safety efficiency

M. F. Makhtar (✉) · H. Albukhari · M. H. Rubani
Traffic Management and Aids to Navigation Division, Malaysia Marine Department, Port Klang, Malaysia
e-mail: fadhli@marine.gov.my

H. Albukhari
e-mail: hairizam@marine.gov.my

M. H. Rubani
e-mail: mhisham@marine.gov.my

Y. Mohd. Hasni
Safety of Navigation and Maritime Communication Division, Malaysia Marine Department, Port Klang, Malaysia
e-mail: yasmin@marine.gov.my

A. Ismail et al. (eds.), *Advancements in Industry and Technology*,
SpringerBriefs in Applied Sciences and Technology,
https://doi.org/10.1007/978-3-031-97931-6_2

1 Introduction

The SOM is a vital international maritime route, one of the world's busiest shipping lanes, and a strategically significant waterway [1]. The secure passage of ships in this area is of utmost significance by providing visible and aural indicators to direct vessels and warn them of possible hazards. Navigational aids (NAVAIDs) are a vital part in assuring the safety and effectiveness of maritime traffic.

In recent years, the advancements in technology and the increasing demand for efficient maritime navigation have prompted the modernization and digitalization of the marine NAVAIDs system in the SOM. This paper aims to explore the aspects of the digitalization process, focusing on integrating digital technologies and enhancing monitoring capabilities for AtoN in the SOM.

The digitalization efforts encompass the adoption of state-of-the-art systems such as the Global Positioning System (GPS), radar, and remote sensing technologies to improve the accuracy and reliability of AtoN data. Additionally, the introduction of advanced communication systems facilitates real-time monitoring and remote control of AtoN, enabling more efficient maintenance and response to navigational issues.

Furthermore, the digitalization of AtoN data allows for developing sophisticated decision support systems and integrating predictive analytics to enhance situational awareness and improve navigational safety. The utilization of big data analytics techniques enables the analysis of historical data and the identification of patterns, contributing to the development of proactive measures for optimizing navigation in the SOM and Malaysian waters.

2 Straits of Malacca

The SOM holds immense strategic importance as a vital international shipping lane, connecting the Indian Ocean and the Pacific Ocean. Geographically, the straits are situated between the Malay Peninsular and the Indonesian island of Sumatra, forming narrow and heavily trafficked waterways [2]. The length of approximately 550 miles (890 km) and a width varying from 1.7 to 55 miles (2.7–89 km), it serves as a critical gateway for maritime transportation, facilitating the movement of goods, energy resources, and international trade [3].

The SOM has a rich historical legacy dating back centuries. It has served as a major trade route since ancient times, connecting the maritime silk route from China to India, the Middle East, and Europe. Several empires and kingdoms, notably Srivijaya, Majapahit, and the Malacca Sultanate, were drawn to the area by its strategic position and aspired to monopolize trade in the area [1]. European colonial powers, such as the Portuguese, Dutch, and British, also recognized the significance of the Straits and established their dominance over the region. The historical developments in the

Straits have shaped the economic and cultural landscape of Southeast Asia and left a lasting impact on the trade dynamic of the world.

The historical developments in the Straits have shaped the economic and cultural landscape of Southeast Asia and left a lasting impact on the trade dynamic of the world. In the SOMS, there is a lot of maritime traffic. Approximately 300 ships travel through the Straits each day, arriving or departing once every two to three minutes [4–6]. There may be about 100,000 vessels passing through this waterway each year [7]. Due to this circumstance, the SOM has turned into a bottleneck for ships transporting essential goods between 600 ports spread over 120 nations [5, 7, 8].

The SOM plays an essential role in driving economic growth in the region. The maritime activities conducted in these waterways, including shipping, logistics, and port operations, generate significant employment opportunities and contribute to gross domestic product (GDP) growth [9, 10]. The presence of major ports and industrial zones along the Straits, such as Port Klang, has attracted foreign direct investment (FDI) and stimulated industrial development, including manufacturing, refining, and logistics industries. The growth of these industries has bolstered economic output, generated revenue, and fostered technological advancement, further enhancing the region's economic prosperity [5, 10].

The economies of the country bordering the Straits of Malacca have greatly benefitted from their proximity and access to these strategic waterways. Singapore has emerged as a global financial and maritime hub, leveraging its location and excellent port infrastructure to become a center for trade, finance, and logistics [5]. The country's development as a premier global city-state owes much to its advantageous position along the Straits. Additionally, Malaysia and Indonesia, which border the Straits, have experienced significant economic development through increased trade, industrialization, and infrastructure projects [5]. The Straits have played an important role in connecting this nation with the global market, facilitating regional integration, and promoting economic cooperation among countries in Southeast Asia [11]. These waterways are used for the bulk of seaborne trade across the world, including the transportation of raw materials, manufactured goods, and energy supplies. The efficient and reliable shipping route provided by the Straits facilitates just-in-time (JIT) delivery, reduces transportation costs, and enables seamless global supply chains (SC).

3 Navigational Aids

Marine navigational aids (NAVAIDs) are essential for enabling safe vessel passage in a variety of waterways, including ports, channels, and coastal regions. When the SOM encounters a high volume of maritime traffic, such as huge container ships, oil tankers, and other vessels, it is crucial in guaranteeing the safe passage of vessels across the Straits [6, 12].

The variety of visible, auditory, and electronic aids that are put in strategic locations to give mariners direction, warning, and positional references. Marine NAVAIDs are navigation aids created specifically to facilitate vessels locate themselves, keep a safe course, and avoid nautical hazards.

They encompass a diverse array of physical structures, such as buoys, beacons, lighthouses, daymarks, radio beacons, and electronic aids [13]. To guide sailors through channels, indicate safe passage, alert them to potential risks, and facilitate position fixing, the aids are placed strategically along waterways. It can be visible, aural, or technological [14].

Marine NAVAIDs play a critical role in maritime navigation by giving seafarers access to vital knowledge and resources that help them navigate properly and securely, especially in difficult terrain and bad weather [14, 15]. NAVAIDs act as trustworthy markers for locating dangerous locations, impediments, and passable water channels, ensuring that ships stay on course and preventing collisions, groundings, or other mishaps [15]. In crowded waterways, NAVAIDs are essential for improving situational awareness, lowering navigational danger, and ensuring efficient vessel movements [16].

The efficiency and potential of maritime NAVAIDs have been considerably increased by technological advancement. The accuracy, dependability, and effectiveness of NAVAIDs have been greatly increased thanks to these technical developments, which have increased their importance for assuring safe navigation preventing vessel congestion [17, 18]. Moreover, it also enhances the safety of vessel operations by warning of navigational hazards, indicating safe routes, and providing timely information on changing conditions. NAVAIDs monitoring system also contributes to the reduction of accidents, grounding incidents, and maritime pollution, thereby minimizing the associated economic and environmental impacts.

4 E-NAV System

E-navigation is the "harmony of marine navigation systems and supporting shore services driven by user needs," according to the International Association of Maritime Aids to Navigation and Lighthouse Authorities (IALA) [13, 18]. By gathering, integrating, exchanging, presenting, and analyzing maritime data electronically, e-navigation improves safety and security at sea and safeguarding the marine environment.

The E-NAV monitoring system developed by the Malaysia Marine Department focusing to monitor the safety of navigation in SOM and Malaysian waters significantly contributes to enhancing maritime safety and lowering the likelihood of maritime accidents and potential environmental harm. The system will arrange the usage of new technologies and guarantee that their use is compatible with the existing electronic navigational and communication technologies and services [17]. It enhances productivity and lowers costs for marine transportation by streamlining the electronic management and information [19].

The system encompasses several key elements to enhancing navigational safety. The system provides electronic navigational charts and integrates real-time data

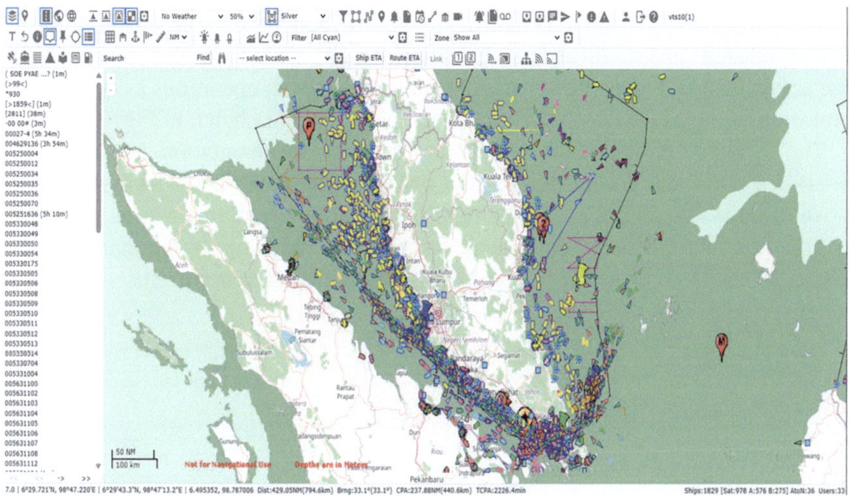

Fig. 1 Real-time vessel monitoring in E-NAV system

onto a digital display as shown in Fig. 1. The E-NAV system incorporates a weather information system, offering meteorological and oceanographic data. An integrated communication system facilitates seamless information exchange among vessels, authorities, and stakeholders [18]. The system enhances situational awareness, enables efficient traffic management, and supports informed decision-making for safer navigation in the SOM and Malaysian waters.

The system integrates multiple sensors to gather real-time data on the condition of AtoN where the sensor is equipped with cameras, a radar system, AIS receiver, and remote monitoring sensor. The sensors continuously collect and store information such as position, status, light characteristics, and environmental conditions. The collected data from the sensor are transmitted to a centralized data management system involving wireless communication networks to ensure a reliable and timely transfer of information and handle a large volume of data from multiple sensors across a wide area.

The system employs advanced data analysis techniques to process the incoming data. Algorithms and software applications are used to interpret sensor data, identify patterns, and detect anomalies. The analysis includes evaluating the position accuracy, determining the health of the AtoN, monitoring the changes in light characteristics, and assessing the impact of environmental factors. The system was designed to monitor and continuously analyze the AtoN data and able to identify any incidents or deviations from expected norms. It could identify and notify any error or abnormal activity such as malfunctioning lights, misalignment, unauthorized tampering, or any abnormalities [20].

Figure 2 shows the E-NAV system detects the incidents that occurred and generates automated alerts and notifications to authorities responsible for maintenance, corrective action, and fast response for search and rescue (SAR) operations as occurred on explosion of the MT Pablo in South China Sea in 2023 as shown in Fig. 3.

The system facilitates incident reporting and documentation by generating detailed reports and analyzing the data based on the historical tracking of vessel movements as shown in Fig. 4. The reports contain relevant information, including the nature of the incident, timestamp, location, and any supporting data from the sensor. It benefits the agency to investigate the accident that occurred.

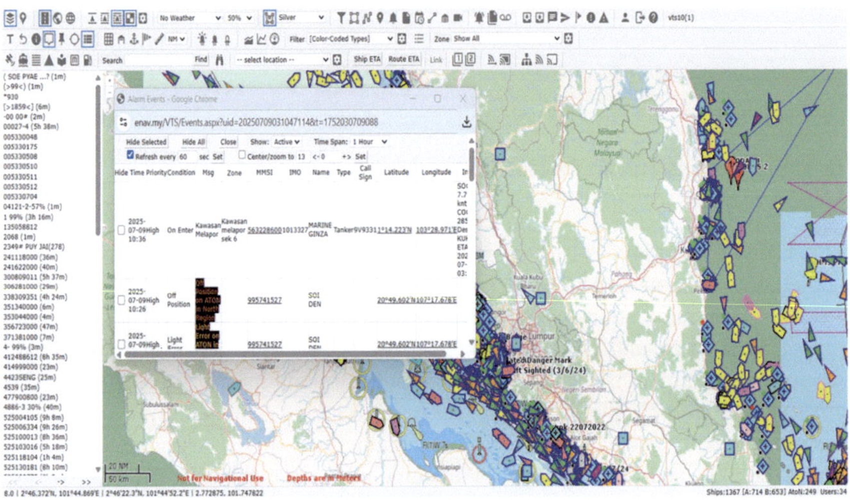

Fig. 2 Real-time incident alert system

Fig. 3 MT Pablo explodes in South China Sea in 2023

Fig. 4 Incident investigation and analysis in E-NAV system

5 Conclusion

The E-NAV system effort would lessen navigational mistakes including maritime accidents and ship-sourced marine pollution in the SOM and Malaysian waters. It shouldn't be seen as a substitute for conventional navigational aids. Since electronic location signals are not error-free, e-navigation should instead include conventional AtoN. The optimal strategy may involve combining traditional navigational tools with electronic navigation. The paper concludes that the digitalization of the marine NAVAIDs monitoring systems can significantly enhance navigational safety and efficiency [17]. The integration of advanced technologies improved the accuracy and reliability of AtoN data. The system permits real-time monitoring and remote control of AtoN, enabling quick repair and efficient navigational response. The digitalization of AtoN data opens avenues for the development of decision support systems and predictive analytics, empowering stakeholders with enhanced situational awareness and risk management capabilities [20]. The paper highlights the potential of the advancement in enhancing the safety, efficiency, and suitability of maritime navigational in SOM, providing valuable insights for further developments in the field of maritime traffic management and navigational system.

Acknowledgements The authors want to express their gratitude to Malaysia Marine Department, Light Dues Board and all the anonymous reviewers for their insightful criticism, funds and useful recommendations that helped to elevate the caliber of this work.

References

1. Z. Sun, Regulation of shipping in the straits of Malacca and Singapore. Cent. Int. Law (2017).
2. M.I.M. Zaideen, J. Jeevan, S.A. Hamid, A.H. Saharuddin, Strategies to promote navigational safety and marine environmental protection: with reference to the Straits of Malacca and Singapore safety. Int. J. e-Navigation Marit. Econ. **11**, 44–053 (2019)
3. N.A. Osnin, Malaysian seafarers: the need for policy review. Marit. Econ. Logist. **6**(4), 360–367 (2004)
4. M. Malaysia Marine Department, Statistik pelaporan kapal ke VTS (2023)
5. X. Qu, Q. Meng, The economic importance of the Straits of Malacca and Singapore: an extreme-scenario analysis. Transp. Res. Part E Logist. Transp. Rev. **48**(1), 258–265 (2012)
6. X. Qu, Q. Meng, Development and applications of a simulation model for vessels in the Singapore Straits. Expert Syst. Appl. **39**(9), 8430–8438 (2012)
7. L. Zhang, Q. Meng, T. Fang Fwa, Big AIS data based spatial-temporal analyses of ship traffic in Singapore port waters. Transp. Res. Part E Logist. Transp. Rev. **129**, 287–304 (2019)
8. D.W. Handani, I.M. Ariana, A.A.B. Dinariyana Malacca and Singapore (SOMS) influenced by Indonesian port development, no. Mastic (2018), pp 365–375
9. P. Economic Planning Unit, The Malaysian economy in figures 2021 (2021)
10. I. Legorburu, K.R. Johnson, S.A. Kerr, Shipping: shipbuilding and maritime transportation (2018)
11. T.P. Minister et al, The ASEAN declaration (Bangkok Declaration) (1967), pp. 4–6
12. A. Mcilgorm, The economic contribution of the marine economy: Southeast Asia leads the way. Trop. Coasts **16**(1), 1–81 (2009)
13. AISM-IALA (2010) Aids to navigation manual NAVGUIDE 2010, reproduction (2010), p. 212
14. N.A. Osnin, N.S.F.A. Rahman, A.F.A. Fuad, Study of navigation system at Kenyir Lake to safeguard societal wellbeing. J. Sustain. Sci. Manag. **2018**(Special Issue 5) (2018)
15. M.I. Hasbullah, N.A. Osnin, N.H. Mohd Salleh, A systematic review and meta-analysis on the development of aids to navigation. Aust. J. Marit. Ocean Aff. 1–21 (2022)
16. U. Nations, Guidelines for the harmonization of navigation rules and regulations. Econ. Soc. Comm. Asia Pacific **1** (2001)
17. C.L. Ping, P.G. Jun, Z.X. Gu, The application and research of navigation-aids inspection and maintenance based on video surveillance. Procedia Eng. **15**, 3088–3092 (2011)
18. S. Bauk, N. Kapidani, Improving safety of navigation by implementing VTS/VTMIS: experiences from Montenegro. Inf. Commun. Environ. (November 2020), 253–258 (2015)
19. N. Kapidani, S. Bauk, I.E. Davidson, Digitalization in developing maritime business environments towards ensuring sustainability. Sustain **12**(21), 1–17 (2020)
20. S. Andrés, F. Piniella, Aids to navigation systems on inland waterways as an element of competitiveness in Ulcv traffic. Int. J. Traffic Transp. Eng. **7**(1), 1–18 (2017)

Impact of Socio-cultural Aspects and Their Role in Influencing Communication Abilities of Students

Jyotsna Saini, Manoj Kumar, Umesh Kumar, Rishabh Shrivastava Ronin, and Jay Prakash Vishwakarma

Abstract Language and communication skills are vital to our personal and professional growth and English, as a global language, has become a daily component of our personal and professional lives. In today's work environment, individuals must have fluency in English. However, mastery in English requires understanding of all four language skills: listening, speaking, reading, and writing. To determine the impact of socio-cultural aspects and their role in influencing the communication abilities of students, the researcher surveyed 453 students of three different universities. A questionnaire was created and distributed. The demographic data were analyzed using frequency counts and percentages, while the study hypotheses were tested using the t-test and Chi-square test. The results emphasize the diverse influences on communication skills, underscoring the role of family dynamics and educational practices in shaping English proficiency among individuals.

Keywords English · Socio-cultural aspects · Communication abilities · Student · Demographic data

---

J. Saini · M. Kumar
Amity University Rajasthan, Jaipur, Rajasthan, India
e-mail: jyotsnasaini2014@gmail.com

M. Kumar
e-mail: mkumar1@jpr.amity.edu

U. Kumar (✉)
Amity School of Engineering and Technology, Amity University Patna, Patna, India
e-mail: umeshkumar2517@gmail.com

R. S. Ronin
Suresh Gyan Vihar University, Rajasthan, India
e-mail: rishabh.shrivastava@mygyanvihar.com

J. P. Vishwakarma
Noida International University, Grate Noida, India
e-mail: jayprakashbhu20@gmail.com

A. Ismail et al. (eds.), *Advancements in Industry and Technology*,
SpringerBriefs in Applied Sciences and Technology,
https://doi.org/10.1007/978-3-031-97931-6_3

1 Introduction

The history of communication itself dates back to the origin of speech around 100,000 BCE [1–3]. The first types of symbols (that we know of) were pictograms, petroglyphs, paintings, and rock carvings or petroglyphs, dating as far back as 50,000 years [4–7]. On the other hand, according to numerous sources (e.g., [8–11]) traders and merchants were responsible for spreading writing to remote areas all across the globe. English is known as "lingua franca," which when translated, means the bridging language. There were around 1.35 billion English-speaking people on the earth as of the end of 2021, and this figure is rising yearly [12, 13]. Most research-related articles are now published in English, and English is swiftly emerging as the language of science. For this reason, a number of universities all over the globe have made it compulsory for their PhD candidates to submit their theses in English [14–16].

Sociocultural is a combination of two words "socio" and "culture," with "socio" standing for society or social characteristics and "culture" standing for a group of people's common values, traditions, and practices. Generally speaking, "sociocultural" refers to the way in which social and cultural aspects interact and impact one another. Society and culture exert a direct influence on the language we use, whether spoken or written. Studying how our culture and society affects "how students communicate" is really important as many studies indicated a clear correlation between a child's learning capabilities and their home environment [17–22].

2 Methodology

The study involved 453 students from three major universities viz. Suresh Gyan-Vihar University, JECRC University, and Jagannath University. The universities were selected randomly from the North-West region of Jaipur city (Rajasthan, India) contributing to the diversity of perspectives.

Of these 453 students, 241 responded on communication and 212 responded on socio-culture background via an online questionnaire consisting of questions related to respondent's demographic details and of relevant questions pertaining to the theme of study. These included how socio-cultural background affects English communication skills? Who is responsible for the poor communication skills among the students? And what problems are faced by a student in learning English language because of their cultural and social background?

Data was collected via Google forms sent personally by the researcher to the university faculty and students and the purpose of the study was explained to them. The collected data was then analyzed using SPSS (Statistical Package for the Social Sciences). The demographic data were analyzed using frequency counts and percentages, while the study hypotheses were tested using the t-test and Chi-square test.

3 Results and Discussion

In the detailed analysis presented in Table 1, significant variations were observed in various aspects related to English-speaking skills. Only 33% of students demonstrated fluency in English speaking, while 54% struggled with fluency, and 12% could not communicate in English. Reading skills were fluent for 83% of students, with 68% understanding spoken English efficiently. Writing skills showed a range, with 42% writing fluently, 46% with some grammatical errors, and 12% facing challenges due to vocabulary, grammar, and spelling. A significant portion (61%) reported a lack of vocabulary knowledge. Communication challenges were evident, as 26% could not reply in English, and 42% tried but faced obstacles such as unclear pronunciation. Confidence in speaking English was lacking for 33% of students. Despite these challenges, 77% expressed satisfaction with the secondary and senior secondary curriculum. Statistical tests confirmed the significance of these findings, indicating that English-speaking skills among Jaipur university students are notably weak. These results can offer valuable insights for targeted improvements in areas identified as needing enhancement.

The analysis, as presented in Table 2, explores the role of gender in various aspects related to English communication skills. The socio-cultural background's influence on communication skills is significantly associated with gender, with a higher percentage of males (36.9%) than females (17.4%) reporting an impact. Thinking in their own first language before replying is not significantly related to gender difference, but awareness of accountability for poor communication skills differs between genders, with 36.5% of males to 15.7% of females attributing it to students themselves. Gender differences are also clear in the ability to communicate well in English within the family, communicative programs, and participation in group discussions and language laboratory practices. These results bring to light the need for gender-specific intervention to develop communication skills in both males and females.

The analysis presented in Table 3 investigates the role of socio-cultural background in swaying English communication skills. The data reveals that factors such as residing in semi-urban, urban, or even rural areas does not significantly impact the English communication skills of a student. Similarly, thinking in one's first language before responding in English does not show a significant relationship with communication skills in English. Additionally, there is no strong evidence linking various factors, including teachers, students, parents, friends, and even curriculum, to poor communication skills. But an important association is found between English communication within the family, suggesting that family members who can converse in English may confidently influence an individual's English communication skills. Furthermore, learning language skills and practicing the methods in a language laboratory demonstrate a noteworthy relationship with English communication skills, demonstrating that those who engage in such actions may have better communication skills. Similarly, participation in communicative programs like debates, etc.,

Table 1 Assessing level of English speaking among the students (t-test)

Level of English-speaking skills		Count	%	t	df	Sig	Results
Level of speaking skills in English	I can speak English fluently	80	33	– 366.42	240	0.000	Significant
	I can speak English but not fluently	131	54				
	I cannot speak in English	30	12				
	Total	**241**	**100**				
Level of reading skills in English	I can read in English fluently	201	83	– 659.30	240	0.000	Significant
	I cannot read in English fluently	40	17				
	Total	**241**	**100**				
Understand the other person if they speak in English	I understand them completely and easily	164	68	– 520.97	240	0.000	Significant
	I find it difficult to understand them completely, and they might have to repeat their statement a few times	77	32				
	Total	**241**	**100**				
Level of writing skills in English	I can write fluently	101	42	– 350.58	240	0.000	Significant
	I can write but with a few grammatical errors	110	46				
	I find it difficult to write in English because of vocabulary/ grammar and spelling	30	12				
	Total	**241**	**100**				
Satisfied with the curriculum applied in secondary and senior secondary level	Yes	185	77	– 353.98	240	0.000	Significant
	No	27	11				
	Not Really	29	12				
	Total	**241**	**100**				
Lack in vocabulary knowledge	Yes	147	61	– 331.42	240	0.000	Significant
	No	61	25				
	Don't know	33	14				
	Total	**241**	**100**				

(continued)

Table 1 (continued)

Level of English-speaking skills		Count	%	t	df	Sig	Results
Unable to reply in English intended	Yes	63	26	– 283.97	240	0.000	Significant
	No	77	32				
	Sometimes	101	42				
	Total	**241**	**100**				
Have an unclear pronunciation	Yes	62	26	– 282.40	240	0.000	Significant
	No	74	31				
	Sometimes	105	44				
	Total	**241**	**100**				
Find a Lack of confidence in speaking English	Yes	80	33	– 277.08	240	0.000	Significant
	No	72	30				
	Sometimes	89	37				
	Total	**241**	**100**				
Corrected by the teacher when you speak in English	Never	19	8	– 163.30	240	0.000	Significant
	Rarely	53	22				
	Sometimes	77	32				
	Often	23	10				
	Always	69	29				
	Total	**241**	**100**				

and group discussions is also found to be notably connected with improved English communication skills.

These results show the complexity of various factors swaying communication abilities and the role that family and educational institutions communication play in determining English language proficiency.

4 Conclusion

The information presented in the study provides some key conclusions. Firstly, socio-cultural background, including the type of area one resides in, does not appear to significantly impact English communication skills. Additionally, there is no discernible relationship between thinking in one's first language before responding in English and communication proficiency. The study suggests that factors like curriculum, friends, parents, students, and teachers alone may not be primary contributors to poor communication skills among students. However, a notable finding is the significant positive relationship between English communication within the family and individual communication skills. Family members who can communicate in English may positively influence an individual's proficiency. Furthermore,

Table 2 Chi-Square test on reason for suffering in English communication skills with gender (a (alpha) which represents significance level)

Reason for suffering in English communication skills		Gender		Total	Pearson's chi-square	df	Sig	Result
		Male	Female					
The socio-cultural background has affected your English communication skills	Yes	89 (36.9%)	42 (17.4%)	131 (54.3%)	1.616[a]	2	0.045	Significant
	No	47 (19.5%)	26 (10.7%)	73 (30.2%)				
	May be	21 (8.7%)	16 (6.6%)	37 (15.3%)				
Total		**157 (65.1%)**	**84 (34.8%)**	**241 (100%)**				
Think first language or mother tongue before replying in English	Yes	70 (29%)	38 (15.7%)	108 (44.8%)	2.040[a]	2	0.361	Not Significant
	No	36 (14.9%)	25 (10.3%)	61 (25.3%)				
	Sometimes	51 (21.1%)	21 (8.7%)	72 (29.8%)				
Total		**157 (65.1%)**	**84 (34.8%)**	**241 (100%)**				
Responsible for the poor communication skills among the students	Curriculum/ syllabus	8 (3.3%)	3 (1.2%)	11 (4.5%)	7.813[a]	6	0.025	Significant
	Friends	5 (2%)	1 (0.4%)	6 (2.4%)				
	Parents	7 (2.9%)	1 (0.4%)	8 (3.3%)				
	Students and teachers both	22 (9.1%)	10 (4.1%)	32 (13.2%)				
	Student themselves	88 (36.5%)	38 (15.7%)	126 (52.2%)				
	Teachers/ lecturers	27 (11.2%)	23 (9.5%)	50 (20.7%)				
Total		**157 (65.1%)**	**84 (34.8%)**	**241 (100%)**				
Who all can communicate in English in the family	Father	27 (11.2%)	22 (9.1%)	49 (20.3%)	11.497[a]	4	0.022	Significant
	Mother	25 (10.3%)	18 (7.4%)	43 (17.8%)				
	Siblings	52 (21.5%)	32 (13.2%)	84 (34.8%)				
	Grand parents	2 (0.8%)	0 (0%)	2 (0.8%)				
	No one	51 (21.1%)	12 (4.9%)	63 (26.1%)				

(continued)

Table 2 (continued)

Reason for suffering in English communication skills		Gender		Total	Pearson's chi-square	df	Sig	Result
		Male	Female					
Total		**157 (65.1%)**	**84 (34.8%)**	**241 (100%)**				
Ever practiced and learned the language skills in language laboratory or lab	Yes	54 (22.4%)	36 (14.9%)	90 (37.3%)	4.662[a]	2	0.009	Significant
	No	40 (16.5%)	26 (10.7%)	66 (27.3%)				
	I don't know what is language laboratory or lab	63 (26.1%)	22 (9.1%)	85 (35.2%)				
Total		**157 (65.1%)**	**84 (34.8%)**	**241 (100%)**				
Ever participated in group discussion/ debate or any other communicative interaction programs in the university	Yes	63 (26.1%)	44 (18.2%)	107 (44.3%)	5.697[a]	5	0.033	Significant
	No	27 (11.2%)	12 (4.9%)	39 (16.1%)				
	Not interested	13 (5.3%)	7 (2.9%)	20 (8.2%)				
	Do not know about these programs	8 (3.3%)	2 (0.8%)	10 (4.1%)				
	I Never got a chance	23 (9.5%)	6 (2.4%)	29 (12%)				
	I tried to participate but felt hesitated so did not	23 (9.5%)	13 (5.3%)	36 (14.9%)				
Total		**157 (65.1%)**	**84 (34.8%)**	**241 (100%)**				

engaging in language laboratory activities is significantly associated with improved English communication skills, indicating the potential benefits of utilizing these educational resources. Active participation in group discussions, debates, or communicative programs is also linked to enhanced proficiency in English communication. In essence, the conclusions emphasize the diverse influences on communication skills, underscoring the role of family dynamics and educational practices in shaping English proficiency among individuals.

Table 3 Chi-square test on reason of suffering in English communication skill with social background

Reason for suffering in English communication skills		Social background			Total	Pearson's chi-square	df	Sig	Result
		Rural	Semi urban	Urban					
Socio-cultural background has affected your English communication skills	Yes	69 (28.6%)	24 (9.9%)	38 (15.7%)	131 (54.3%)	7.675a	4	0.104	Not significant
	No	28 (11.6%)	14 (5.8%)	31 (12.8%)	73 (30.2%)				
	May be	12 (4.9%)	10 (4.1%)	15 (6.2%)	37 (15.3%)				
Total		**109 (45.2%)**	**48 (19.9%)**	**84 (34.8%)**	**241 (100%)**				
Think first language or mother tongue before replying in English	Yes	55 (22.8%)	23 (9.5%)	30 (12.4%)	108 (44.8%)	7.827a	4	0.098	Not significant
	No	24 (9.9%)	8 (3.3%)	29 (12%)	61 (25.3%)				
	Sometimes	30 (12.4%)	17 (7%)	25 (10.3%)	72 (29.8%)				
Total		**109 (45.2%)**	**48 (19.9%)**	**84 (34.8%)**	**241 (100%)**				
Responsible for the poor communication skills among the students	Curriculum/ Syllabus	3 (1.2%)	3 (1.2%)	5 (2%)	11 (4.5%)	7.413a	12	0.829	Not significant
	Friends	3 (1.2%)	0 (0%)	3 (1.2%)	6 (2.4%)				
	Parents	5 (2%)	1 (0.4%)	2 (0.8%)	8 (3.3%)				
	Students and teachers both	11 (4.5%)	10 (4.1%)	11 (4.5%)	32 (13.2%)				
	Students himself/ herself	55 (22.8%)	22 (9.1%)	41 (17%)	118 (48.9%)				
	Teachers/lecturers	32 (13.2%)	12 (4.9%)	22 (9.1%)	66 (27.3%)				
Total		**109 (45.2%)**	**48 (19.9%)**	**84 (34.8%)**	**241 (100%)**				
Who all can communicate in English in the family	Father	26 (10.7%)	8 (3.3%)	15 (6.2%)	49 (20.3%)	54.578a	8	0.000	Significant
	Mother	9 (3.7%)	9 (3.7%)	25 (10.3%)	43 (17.8%)				
	Siblings	26 (10.7%)	20 (8.2%)	38 (15.7%)	84 (34.8%)				

(continued)

Table 3 (continued)

Reason for suffering in English communication skills		Social background			Total	Pearson's chi-square	df	Sig	Result
		Rural	Semi urban	Urban					
	Grand parents	0 (0%)	2 (0.8%)	0 (0%)	2 (0.8%)				
	No one	48 (19.9%)	9 (3.7%)	6 (2.4%)	63 (26.1%)				
Total		**109 (45.2%)**	**48 (19.9%)**	**84 (34.8%)**	**241 (100%)**				
Ever practiced and learned the language skills in language laboratory or lab	Yes	32 (13.2%)	20 (8.2%)	38 (15.7%)	90 (37.3%)	15.841a	4	0.003	Significant
	No	24 (9.9%)	15 (6.2%)	27 (11.2%)	66 (27.3%)				
	I don't know what is language laboratory or lab	53 (21.9%)	13 (5.3%)	19 (7.8%)	85 (35.2%)				
Total		**109 (45.2%)**	**48 (19.9%)**	**84 (34.8%)**	**241 (100%)**				
Ever participated in group discussion/ debate or any other communicative interaction programs in the university	Yes	40 (16.5%)	24 (9.9%)	43 (17.8%)	107 (44.3%)	21.088a	10	0.020	Significant
	No	23 (9.5%)	9 (3.7%)	7 (2.9%)	39 (16.1%)				
	Not interested	16 (6.6%)	1 (0.4%)	3 (1.2%)	20 (8.2%)				
	Do not know about these programs	6 (2.4%)	1 (0.4%)	3 (1.2%)	10 (4.1%)				
	I Never got a chance	11 (4.5%)	5 (2%)	13 (5.3%)	29 (12%)				
	I tried to participate but felt hesitated so did not	13 (5.3%)	8 (3.3%)	15 (6.2%)	36 (14.9%)				
Total		**109 (45.2%)**	**48 (19.9%)**	**84 (34.8%)**	**241 (100%)**				

References

1. S. Miyagawa, The integration hypothesis of human language evolution and the nature of contemporary languages. Front. Psychol. **5** (2014)
2. J. Baines, The earliest Egyptian writing: development, context, purpose, in *The First Writing: Script Invention as History and Process*, ed. by V. Boudreau (Cambridge University Press, 2004), pp. 150–189
3. R. Panini, F. Fiorini, Oral communication: short history and some rules. G. Ital. Nefrol. **32**(2) (2015)
4. P.M. Lester, *Visual Communication: Images with Messages* (Illustrate). (Thomson Wadsworth, 2006). https://books.google.co.in/books?vid=ISBN0534637205&id=6oibH9roTmkC&pg=PA48&lpg=PA48&dq=oldest+Chauvet&sig=Wavi-vRU4yanySHdKiYPWO70_os&redir_esc=y
5. B. Ferreira, Pig painting may be world's oldest cave art yet, archaeologists say. The New York Times (2021). https://www.nytimes.com/2021/01/15/learning/lesson-of-the-day-pig-painting-may-be-worlds-oldest-cave-art-yet-archaeologists-say.html#:~:text=In%20a%20hidden%20valley%20on,discovering%20and%20analyzing%20the%20painting.
6. A. Brumm, Oldest cave art found in Sulawesi. Sci. Adv. **7**(3), Jan. eabd4648 (2021)
7. L. Nemo, Petroglyphs in the U.S.: What native communities want you to know about these rock carvings. Discov. Mag. (2021). https://www.discovermagazine.com/environment/petroglyphs-in-the-u-s-what-native-communities-want-you-to-know-about-these
8. J.J. Mark, *Writing—World History Encyclopedia*. (World History Publishing, 2022). https://www.worldhistory.org/writing/
9. I. Regulski, *The Origins and Early Development of Writing in Egypt*, vol. 1. (Oxford University Press, 2016). http://oxfordhandbooks.com/view/https://doi.org/10.1093/oxfordhb/9780199935413.001.0001/oxfordhb-9780199935413-e-61
10. P.T. Daniels, The first civilizations, in *The World's Writing Systems*, ed. by P.T. Daniels, W. Bright (1996), p. 24
11. S. Chrisomalis, The origins and co-evaluation of literacy and numeracy, in *The Cambridge Handbook of Literacy*, ed. by D.R. Olson, N. Torrance (Cambridge, 2009), pp. 62–64
12. EF-EPI, World English proficiency index—executive summary. Educ First (2022). https://www.ef-slovakia.sk/epi/executive-summary/
13. NU.edu, How to improve your communication skills in English. National University Resources (2022). https://www.nu.edu/blog/how-to-improve-communication-skills-in-english/#:~:text=She%20recommends%20listening%20to%20songs,Watch%20news%20programs%20in%20English
14. C. Tardy, The role of English in scientific communication: lingua franca or Tyrannosaurus rex? J. English Acad. Purp. **3**(3), 247–269 (2004)
15. R. Elnathan, English is the language of science—but precision is tough as a non-native speaker. Nature (2021). https://doi.org/10.1038/d41586-021-00899-y
16. British Study Centres, The importance of the English Language: 10 reasons to learn English. British Study Centres (2022). https://web.archive.org/web/20220314061450/, https://www.british-study.com/en/importance-of-english-language
17. L. Woodrow, Anxiety and speaking English as a second language. RELC J. **37**(3), 308–328 (2006)
18. L.A. Kurdek, R.J. Sinclair, Relation of eighth graders' family structure, gender, and family environment with academic performance and school behavior. J. Educ. Psychol. **80**(1), 90–94 (1988)
19. K. Marjoribanks, Environment, social class, and mental abilities. J. Educ. Psychol. **63**(2), 103–109 (1972)
20. T. Kellaghan, Relationships between home environment and scholastic behavior in a disadvantaged population. J. Educ. Psychol. **69**(6), 754–760 (1977)

21. A. Burns, R. Homel, Social inequalities and adjustment to school. Aust. J. Educ. **29**(1), 76–91 (1985)
22. S.I. Khan, Living on the extreme margin: social exclusion of the transgender population (Hijra) in Bangladesh. J. Heal. Popul. Nutr. **27**(4), 441–451 (2009)

Investigation of Radiation Attenuation of Co-shielding Materials for Pipe Inspection Gauge Application

Noraishah Othman, Ahmad Zubair Johan, and Muhammad Anas Mat Marzuki

Abstract Co-shielding is adopted in this study as preliminary investigation in order to assess the potential of a radioactive pigging system in leak detection application. Pipe inspection gauge (PIG) is the insertion of an encapsulated system inside the underground pipeline to detect the leakage along the pipe under investigation. The developed PIG comprises of radiation sensor, data logger, and battery to power up the system. The radioactive source is injected inside the fully occupied pipeline and seeps through any cracks and stays outside of the pipeline. The launched PIG picks up the radioactive signals that signifies the location of the leak. Normally, the encapsulated system is made of stainless steel or aluminum material. A 25.4-cm mild steel pipeline is set as pipe under investigation, and the distance between PIG and pipe is set as 150 cm. In this paper, a desktop experiment is undertaken and discussed in order to optimize the gamma intensity received by the radiation sensor, the thickness of the material, and the type of co-shielding materials, respectively. A desktop experiment is a simple radioactive set up to describe the gamma attenuation effect with different configurations. Sodium iodide (NaI) as scintillation detector is used as the radiation sensor and Cs-137 (energy of 0.662 MeV and half-life of 30 years) with activities of 74 MBq and 740 MBq as low and high activities are adopted in this study. Moreover, mild steel–aluminum (MS-Al) and mild steel–stainless steel (MS-SS) with different configurations are used as co-shielding material for gamma attenuation study.

Keywords Co-shielding material · PIG system · Gamma attenuation · Desktop experiment · Cs-137

N. Othman (✉) · A. Z. Johan · M. A. Mat Marzuki
Industrial Technology Division, Malaysian Nuclear Agency, Kajang, Selangor, Malaysia
e-mail: noraishah@nm.gov.my

A. Z. Johan
e-mail: zubair@nm.gov.my

M. A. Mat Marzuki
e-mail: anas@nm.gov.my

A. Ismail et al. (eds.), *Advancements in Industry and Technology*,
SpringerBriefs in Applied Sciences and Technology,
https://doi.org/10.1007/978-3-031-97931-6_4

1 Introduction

A study of photon attenuation such as X-ray, γ-ray, and bremsstrahlung ranging from 1 keV to 20 MeV has been extensively conducted for 40 elements extending from hydrogen ($Z = I$) to uranium ($Z = 92$) [1]. In order to estimate the gamma transmission counts through a chosen thickness of shielding material or optimize the shielding thickness of material at desired level of attenuation, the attenuation coefficient and the material density are adopted as notable parameters [2]. The attenuation coefficient is often expressed in terms of unit area per mass (cm^2/g) and describes how electronic radiation traverses through matter. Attenuation of well-collimated gamma rays obeys the Beer-Lambert law (Beer's Law), as stated in Eq. (1) which describes the gamma transmission principle which is the intensity of the transmitted beam [3].

$$I = I_o e^{-(\mu/\rho)x}, \tag{1}$$

where I is the intensity after shielding, I_o is the incident intensity, μ is the mass attenuation coefficient (cm^2/g), ρ is the density of the shielding material (g/cm^3), and x is the thickness of the shielding material (cm).

Primary processes occur when a narrow beam of monoenergetic photons is attenuated through a layer of material with mass per unit area which are scattering and absorption in the matter as well as traversing through matter, respectively. The scattering and absorption activities involve when the radiations interact with matter, whereas the transmitted beam indicates no gamma interaction with matter. Thus, the linear attenuation coefficient is the probability of interaction that this photon may have [4]. However, selecting the most suitable shielding material for a given source of ionizing radiation requires understanding of the fundamental principle gamma ray interactions with matter, application of attenuation data from available resources, and last but not least knowledge of the source of radiation. Therefore, the objective of the study is to investigate the radiation attenuation with respect to thickness, variation of Cs-137 activity, and type of co-shielding materials.

2 Methodology

Radiation measurements are performed using a Ludlum model 2200 scaler ratemeter attached to a Ludlum sodium iodide (NaI) scintillation detector. The desktop experiment utilizes a well-collimated point source of gamma radioactivity (Cs-137) in a half-inch lead with a 5-mm aperture and arranged 150 mm away from NaI scintillation detector as shown in Fig. 1. Mild steel–aluminum (MS-Al) sheet and mild steel–stainless steel (MS-SS) are used as co-shielding material and the thickness of the respective materials are shown in Table 1. In this study, 2 inch diameter of NaI detector is used and exposed to low and high activity of Cs-137 which are 74 MBq and 740 MBq, respectively. In order to relate the desktop experiment with PIG system,

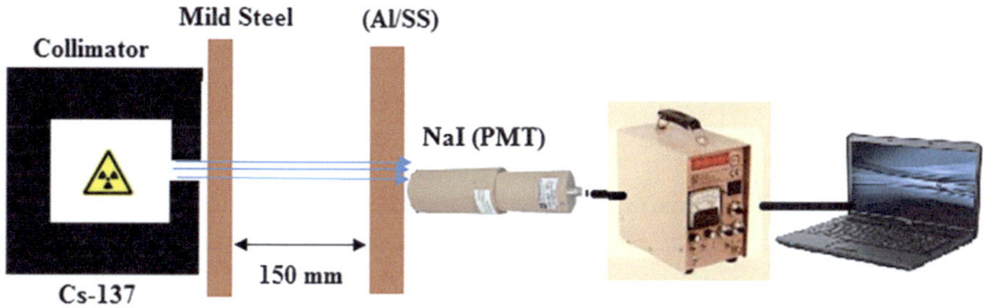

Fig. 1 Configuration of desktop experiment

Cs-137 is located outside of the mild steel (MS) pipeline, NaI detector is kept inside the aluminum (Al) or stainless steel (SS) encapsulated PIG system, and the distant between PIG and radioactive source is set at 150 cm. Example of the PIG system is shown in Fig. 2. However, the PIG experiment is not covered in this paper.

Table 1 Configuration of co-shielding materials

Activity (MBq)	Configurations			
	Label		Thickness (mm)	
	MS-Al	MS-SS	(MS)	(Al)
74 (Low activity)	A	P	10	10
	B	Q	10	20
740 (High activity)	C	R	20	10
	D	S	20	20

Fig. 2 Example of PIG system set up inside the pipeline

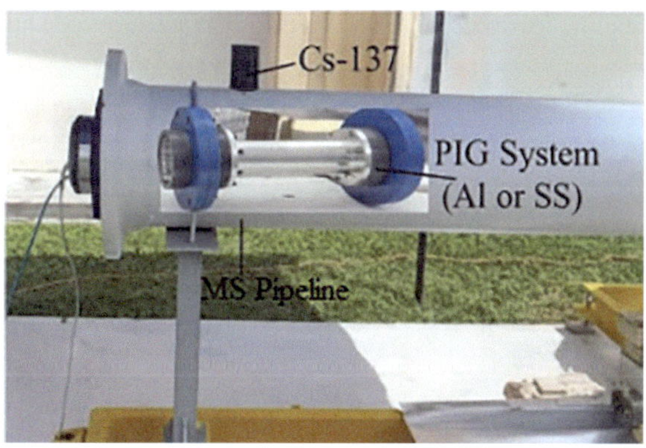

3 Results and Discussion

3.1 *Mild Steel–Aluminum (MS-Al)*

Figure 3 shows the gamma attenuation with respect to configurations when low activity is adopted. It is observed clearly that the radiation counts are attenuated significantly when MS-Al as co-shielding material is exposed to low radiation. Moreover, the radiation counts are highly detectable when passing through the co-shielding material from A to D sequentially. The results show that the detector is able to acquire radiation signals from any configurations and the radiation counts are higher when configuration A is adopted followed by B, C, and D. The results are predictable since the combined thickness of A is 20 mm which is lesser than B, C, and D, respectively. The higher thickness is causing the radiation counts to be reduced as shown by D, C, and B. The results also adhere to Beer–Lambert's principle in which gamma intensity is inversely proportional to material thickness [5, 6]. Moreover, the radiation counts for B is almost comparable with A although the thickness of aluminum (Al) is increased to 20 mm this is due to the thickness of Al is less significant than mild steel (MS) due to its density. The density of MS is 7.85 g/cm^3 a lot higher than Al with 2.70 g/cm^3. Although B and C are having common combined thickness (30 mm), the attenuation difference is very significant due to the thickness of MS is increased to 20 mm and the fact that the density of MS is higher than Al. Therefore, it is inevitable that D provides low intensity due to higher thickness which is 40 mm than other configurations.

However, when air count is measured as reference counts with respect to configurations, it is observed that all the detected counts are higher although the co-shielding materials are assigned between them. That phenomenon is due to the detector has experienced paralyzed dead time although low activity is adopted. It can be implied that the fixed distance of 150 mm between source and detector is in close proximity

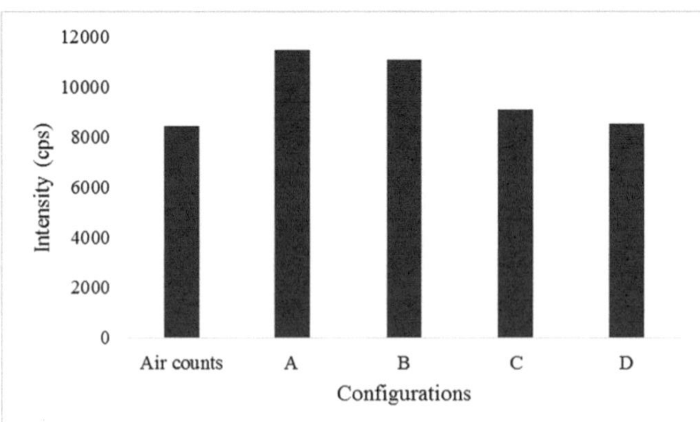

Fig. 3 Radiation attenuation of low activity versus MS-Al configuration

and thus leads to overwhelmed photons to the detector. Hence, intervention of thickness has remediated the situation and led the results to follow the Beers–Lambert's principle.

Figure 4 shows the gamma transmission counts of high activity Cs-137 after traversing the different configuration of MS-Al co-shielding material. However, the recorded gamma counts are very much lower compared to Fig. 3. The results are peculiar since it violates the Beer's Law. However, there is the dead time phenomenon intrinsically experienced by the detector which cannot tolerate the excessive amount of ionizing radiation [7, 8]. Due to avalanche amounts of photons, the detector is overwhelmed by the radiation incident and blind, thus, unable to record the upcoming radiation and lost the data. Therefore, during dead time detection system, any excessive radiation pulses that reach the saturation rate are undetected by the detector and losing significant amount of data that leads to lower radiation counts. Figure 4 also displays the radiation count is lower when high activity is adopted. Hence, air count reading indicates lower counts due to the overzealous of photons existence that causes the paralyzed dead time. Air count is defined as measurement of radioactive intensity without presence of shielding material. In order to alleviate the intensity reading, the shielding thickness is increased gradually from $A = 20$ mm, $B = 30$ cm, $C = 30$ cm, and $D = 40$ cm, respectively. Thus, the dead time phenomena is overcome by increasing the thickness of the shielding material.

Apparently, in Compton scattering experiments, when gamma rays are impinged on a thicker object, the photons experience further interaction inside the geometry of the target material and multiple scattering occurs within the object prior leaving the target with diminished energy. Thus, the recorded intensity increases with increased in target thickness due to multiply backscattered photons occurrence within the target before leaving the target [9, 10]. Some of the photons experience absorption and transmission and detectable by NaI PMT. However, the results also indicate that intensity of A and B is comparable, whereas C and D is almost the same. The effect

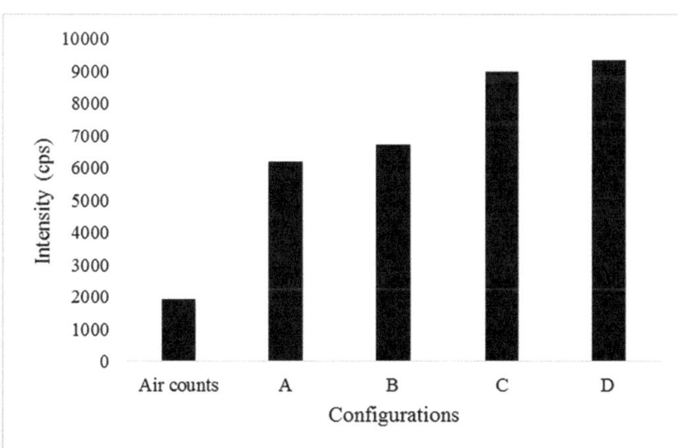

Fig. 4 Radiation attenuation of high activity versus MS-Al configuration

of intensity is affected by the mild steel density and not so much on aluminum since its density is much lower than mild steel. In short, the gamma intensity is in the order of $A > B > C > D$ when high activity is adopted. In conclusion, additional of thickness has improved the gamma intensity during dead time phenomenon.

3.2 Mild Steel–Stainless Steel (MS-SS)

Figure 5 shows that when low activity Cs-137 is exposed to the MS-SS configurations, radiation is attenuated subsequently when passing through low thickness to high thickness target material that corresponds to 20 mm, 30 mm, and 40 mm, respectively. Since gamma intensity is in the order of P > Q > R > S with respect to thickness, the exposure of low activity Cs-137 to MS-SS co-shielding material also obeys the Beer–Lambert's principle. In this section, similar discussion can be found as elaborated by Fig. 3. However, since MS and SS are having comparable density and higher than density of Al, the intensity of gamma receives by each configuration is lower than by MS-Al co-shielding materials.

Figure 6 shows that the gamma intensity after traversing material P is reduced although high activity of Cs-137 is adopted. As discussed in Sect. MS-Al, the occurrence of dead time causes the unrecorded gamma counts when high activity is adopted which causes the low intensity unlike when low activity is adopted. In this arrangement, the fixed distance of 150 mm has also aggravated the received intensity by the detector. Apparently, the distance is insufficient for the detector to process its upcoming incident. Figure 6 also indicates that overwhelming photons have caused P with lowest thickness (i.e., 20 mm) to have reduced intensity compared to others.

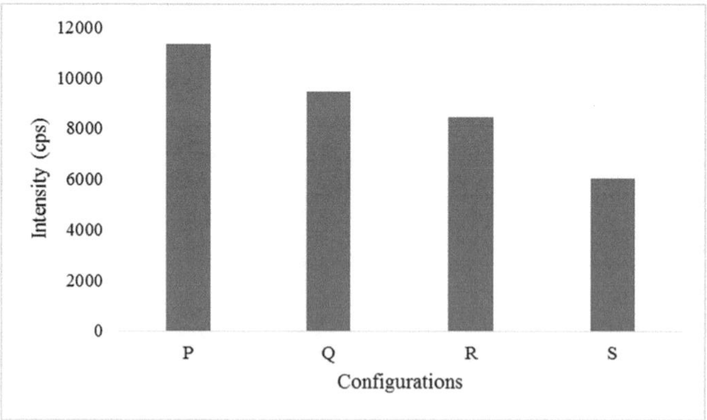

Fig. 5 Radiation attenuation of low activity versus MS-SS configuration

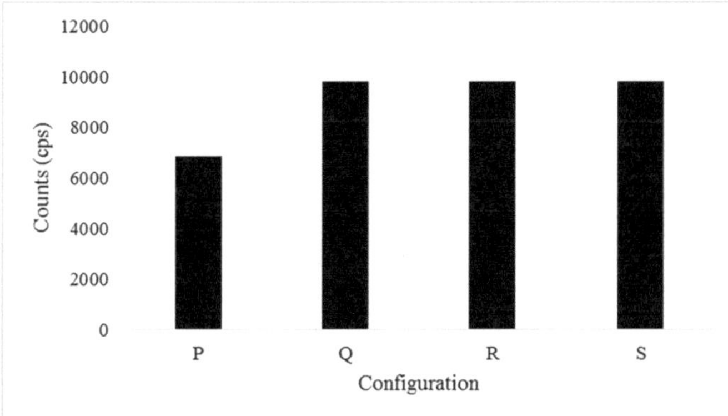

Fig. 6 Radiation attenuation of high activity (740 MBq) versus MS-SS

However, when higher thickness is introduced as shown in R, S, and T, the intensity counts increase that indicate the optimum thickness is achieved for MS-SS co-shielding material. In conclusion, the dead time is reduced at a thickness of 30 mm for MS-SS configuration.

Finally, in order to relate this result with the pigging system, we can optimize the thickness for each pigging capsule and identify allowable thickness of piping system so that the radiation can acquire radiation data accurately. Besides, the selection of material can also be made whether to adopt aluminum or stainless steel capsule for the pigging system. On top of that, the identification of dead time of radiation detector and optimization of suitable radioactive activity are also needed to ensure no signals are lost during data acquisition. Last but not least, the optimum thickness for each co-shielding material has been identified which has assisted in designing the effective pigging system in the future.

4 Conclusion

Several key takeaways can be extracted from this works:

1. Gamma radiation can easily traverse low atomic weight of mild steel–aluminum and mild steel–stainless steel materials irrespective of its gamma intensity (activity) and thickness configuration
2. For both MS-Al and MS-SS co-shielding materials, the Lambert–Beer's principle is adhered to for low activity (74 MBq) Cs-137 in which intensity is inversely proportional to thickness of material
3. For both MS-Al and MS-SS co-shielding materials, the Lambert–Beer's principle is violated for high activity (740 MBq) Cs-137 in which intensity is directly proportional to thickness of material due to dead time phenomenon

4. Dead time phenomenon of radiation detector causes loss of data when high activity Cs-137 is adopted
5. The dead time phenomenon is reduced at thickness of 30 mm that causes the plateau of gamma intensity readings for mild steel–stainless steel configuration
6. Researchers should conduct the dead time study of each detector before embarking any radiation experiment. In this case, identifying the optimum gamma activity will salvage the significant data from losing due to dead time phenomena.

Acknowledgements The authors would like to acknowledge the Ministry of Science Technology and Innovation (MOSTI) with Project Code TDF11211509 that sponsors the project and greatest appreciation to all colleagues from Prototype and Design Centre (PDC) and Plant Assessment Technology (PAT) from Malaysian Nuclear Agency who have directly and indirectly participated in this work.

References

1. J.H. Hubbell, Photon mass attenuation and energy-absorption coefficients from 1 KeV to 20 MeV. Int. J. Appl. Radiat. Isot. **33**, 1269–1290 (1982)
2. R.M. Daniel, Gamma ray attenuation properties of common shielding materials. Dissertation, University Lane Lisle (2018)
3. G.M. dos Alessandra, S.F.D. de Roos, L.S. William, M.S. César, Determination of mass attenuation coefficient in aluminum alloys using code MCNP6, in *Proceedings of the ISSSD*, vol. 1 (2021)
4. M.E. Turgay, Comparison of gamma sources (cobalt and cesium) for density measurement of metals and alloys via using transmission technique. J. Tech. Appl. Sci. **6**, 37–43 (2021)
5. F.R.Ningsih, W.N.Gama, W. Wibisono, A quantitative approach to investigate downpour condition of distillation column using gamma beam radiation. J. Phys. Conf. Ser. 1511, 012078 (2020). https://doi.org/10.1088/1742-6596/1511/1/012078
6. W. Qiming, M. Yugao, Gamma ray thickness measurement. IOP Conf. Ser. Earth Environ. Sci. **237**, 032072 (2019)
7. Patil, A.: Dead time and count loss determination for radiation detection systems in high count rate applications. Dissertations, Missouri University of Science & Technology (2010)
8. U. Shoaib, P. Amol, Radiation detector dead time and pile up: a review of the status of science. Nucl. Eng. Technol. **50**, 1006–1016 (2018)
9. D.S. Arvind, S. Manpreet, S. Bhajan, B.S. Sandhu, Response function of NaI (Tl) detectors and multiple backscattering of gamma rays in aluminum. Appl. Radiat. Isot. **66**, 1467–1473 (2008)
10. D.S. Arvind, B.S. Sandhu, S. Bhajan, Investigations of energy dependence of saturation thickness of multiply backscattered gamma photons in elements and alloys-an inverse matrix approach, International Nuclear Physics Conference 2010 (INPC2010). J. Phys. Conf. Ser. **312**, 052021 (2011). https://doi.org/10.1088/1742-6596/312/5/052021

The Rise of Behavioral Dimension in Social Sciences

Smita Tiwary Ojha, Smita Singh, Kumar Anand, and Ankit Srivastava

Abstract The ideas of behavioral dimension of interaction and social relation have been among the core categories through which many contemporary debates visualize or speculate in institutional setting. So, this paper attempts to study behavior plays in interactions for the sake of good governance and prosperity of an organization. In the popular understanding, behavior science is a science which studies behavioral interactions of individuals and animals too but has limited appeal in wider social science. Since the methods used in behavior process are naturalistic observations, controlled scientific experimentation and mathematical modeling, it also involves rigorous formulations to reach any conclusion. That is why psychology, psychobiology, and cognitive science are integral parts of behavioral science. Furthermore, sociology and behavioral economics also play a significant role in understanding dynamics of the individual at all levels including economics and finance. The article discusses the various facets of behavioral science and its applicability to various streams of social science in India by analyzing different articles published in journals.

Keywords Behavioral modeling · Cognition · Intuition · Microsimulation · Resilience · Rational behavior

---

S. T. Ojha (✉) · K. Anand
Amity University Noida, Noida, India
e-mail: stojha@amity.edu

K. Anand
e-mail: kanand@ptn.amity.edu

S. Singh
Amity University Patna, Patna, India
e-mail: ssingh1@ptn.amity.edu

A. Srivastava
Study Hall College, Lucknow, India
e-mail: ankkit1205@gmail.com

A. Ismail et al. (eds.), *Advancements in Industry and Technology*,
SpringerBriefs in Applied Sciences and Technology,
https://doi.org/10.1007/978-3-031-97931-6_5

1 Introduction

Behavioral sciences study how we communicate in relationships at personal or professional levels. It deals with cognitive entities and relational dynamics in a social system. It deals with fields like cognitive psychology, organizational behavior, sociology as in social psychology, social networks, economics, and finance by using heuristics and microsimulation which is an agent-based model. Microsimulation is a group of computerized analytical tools that inspects activities of individuals. For example, financial transactions, pathogen, spreading disease through a population.

So, it deals with rules-based structures. It can be seen in a traffic microsimulation model where effectiveness of lengthening a turn lane add an intersection is evaluated to decide whether it is worth of spending money on lengthening the lane as the parameters depend upon individual preferences and tendencies. For example, some drivers can wait for a large gap before turning while others always remain in a hurry and accept small gaps. Similarly, in a public health model everyone has different resistance levels and also different personal habits affect transmission of virus. Here modeling techniques play an important role.

Hence, behavior science plays a role in all the aspects of life at personal, psychological, organizational, social and financial level. For understanding behavior science, each entity should be analyzed, so that a complete framework of the individual at personal and professional level can sustain.

2 Methodology

This paper adopts qualitative methods and an interpretive approach to discern the rise of behavioral dimension across social science enquiry. The research includes an analysis of various theoretical contributions being made in the relevant realm to explain social phenomena. Clearly, this implies the use of secondary sources of data. In order to judge the validity of the conclusions drawn in the previous research, attempt is made to apply the same principle to real-life scenarios in current context in India.

3 Results and Discussion

As man is a social animal and animals being tamed by humans, both are integral part of the society. Therefore, understanding of interactional behavior of both is very important for the proper functioning of the society. Humans as an individual work in various areas of society at individual and professional levels [1]. Hence, proper training and stimulation are much needed for them to prosper in every walk of life with resilience. Resilience is the power to bounce back from a difficult situation either at physical, psychological, financial, or social levels. Besides being tamed at home, animals also work in the military and other professional levels too. Hence, animals also require training in behavior science.

There are two categories in behavioral science. (1) Information processing sciences and (2) Social-relational sciences.

Information processing sciences pertain to the process of stimuli received by cognitive entities like sensation—it is related to our sense organs, perception—it is related to how we are perceiving the signals grasped by our sense organs, personality. This component of the individual is related to the genetic and environmental influences. Intelligence—it is global capacity of the individual to think rationally, act purposefully and deal effectively in an environment in all aspects. Attitude—it is the settled way of thinking about others influenced by past experience and knowledge. Thinking—it is the process of reasoning about something and decision-making, including problem solving. These entities help in further decision-making in any situation for the survival in the social environment. Psychology, social cognition, social psychology neuroscience neural networks cognitive science behavior analysis and management are part of this category.

As an integral part of psychology and organizational behavior, study of behaviorism thus becomes mandatory as a part of soft skill training. It is a direct link between different aspects of psychology and human behavior. Figure 1 showcases the different aspects of behavioral science as applied to psychology.

Economics is one of the youngest social sciences in terms of modern method applications. This is why it is a pioneer in adopting new methods and incorporating advances in different fields as part of economics research methods. The increasing use of behaviorism is one more example of this incorporation. It combines psychology with economics to come up with explanations regarding human behavior, mainly decision-making, in the realm of markets and is called behavioral economics.

Behavioral economics with the concept of *bounded rationality* [2, 3] suggests that human behavior depends on cognitive biases, emotions, social influences, time, and space. These lead to people making decisions without going in for the deliberate, controlled, and probabilistic processes as suggested by traditional economic thinking. In 1981, Tversky and Kahneman showed that when choosing between gain and loss

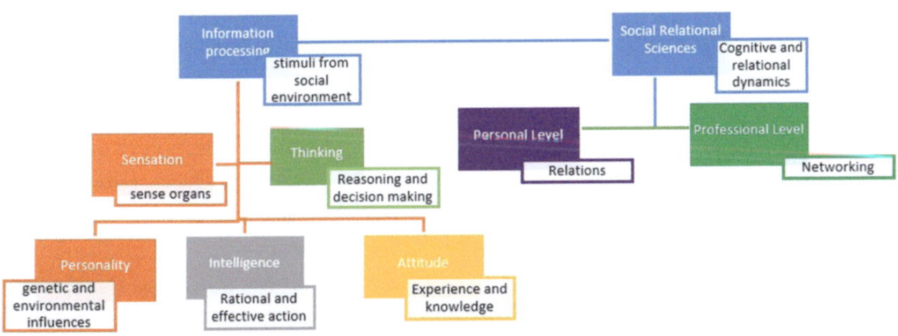

Fig. 1 Behavioral science as applicable to psychology and organizational behavior

options, actual human selections were contrary to those suggested by pragmatic economic theory [4]. They said that a larger proportion of people tend to opt for a riskless alternative when available choices relate to gain, and people are more likely to choose the riskier option if the alternatives are phrased as losses. This idea is often discussed as the *prospect theory* which says that people "dislike losses" more than they like an equivalent gain. "Giving something up is more painful than the pleasure we derive from receiving it." Kahneman and Tversky successfully pointed out that decisions regarding risks are influenced by the way the choices are framed, i.e., they are context dependent [4].

Another interesting behavioristic theory pertains to "*heuristics and biases.*" Goldstein and Geiringer have proposed that instead of applying lengthy and complex mathematical relations, people often choose simple mental "shortcuts" to reach decisions that give near-optimal results in the face of limited information [5].

Richard Thaler, one of the most famous proponents of behavioral economics coined the term *mental accounting.* He held that instead of opportunity costs or trade-offs, people are more prone to concentrate on the sink-cost fallacy. Through his research in the 1990s with Johnson he pointed out that money's origin and planned use determined its utility instead of the "bottom line" [6]. The fungibility of money is another factor determining the expenditure policy adopted by people. This makes them view some returns/gain were by "disposable *house money.*" The tendency of even seasoned investors to consider each mental account pertaining to individual investments separately instead of looking at the whole portfolio has been explained by these economists in terms of behavioralist [7]. Another example of behavioral economics in action is reflected in the increased use of plastic money instead of cash in modern times. Zallermayer said that the *pain of paying* is more pronounced in the case of cash transactions as compared to credit card purchases as the depletion of monetary resources is less clear in the latter case along with a chance of deferred payment [8].

Schwartz has associated *over choice* or choice overload with unhappiness [9]. He held that the large plethora of choices available today in various consumption goods is an important reason behind the youth feeling overwhelmed. Iyengar and Lepper and Iyengar in 2000 showed that decision fatigue, choice deferral and going with the default option were all defense mechanisms adopted by people when faced by too many decisions in the light of little or no information and time constraints [10]. The general tendency to go with the default option to overcome the information overload and an avoidance tendency regarding accountability for decisions also induces consumers to go for default options. This is generally observed even in everyday decisions such as purchase of consumer goods and computers etc. which have several add-ons or variants making it difficult for potential buyers to differentiate among the choices available. It has been observed that instead of following the rational of the "economic man" to maximize returns, buyers simply opt for the default option. In *Nudge*, Thaler and Sunstein point out that experience good information and prompt feedback can be key factors enabling people in making good decisions especially for environmental behavior, health, etcetera [6]. Feedback can be useful in overcoming limited information which causes bounded rationality. Dan Ariely in *Predictably irrational* demonstrated that firms can increase sales by simply highlighting an earlier larger price in comparison with current lower one [11]. The idea of getting the same commodity at a lower price entices consumers to buy goods. Ariely even suggested that free products are considered as more valuable by consumers when sold as a set as seen in increasing sales of commodities sold together.

Financial behavior can be termed as a tool which does not govern only individual or personal attitude toward finance, however, how in totality it can affect. It decides how finance and funds are being managed, how savings or investment decisions are taken, what would be financial plans and goals. The various studies and research *being conducted shows that financial behavior is very effective on financial satisfaction. Financial satisfaction can be understood as a level achieved by an individual that shows the level of welfare regarding current financial condition, that is measured by the existence of a healthy, happy, and financial worry free.* Common financial behavior involves cash, credit, and saving behavior. The people who possess knowledge of finance management have shown a positive trend of more savings and investing for the future, save themselves by resorting to excessive debt which can also put them in financial trouble due to debt trap [12]. Figure 2 highlights the various aspects of behaviorism as applied to finance.

As per the recent surveys and studies performed by few esteemed financial organizations for India, approx. 60% or above of the population is financially insecure, approx. 20% or more of the population is investing in stocks, mutual funds, ULIP, and physical assets. Hence, it is very necessary for the government as well as other concerned financial institutions to pay special attention toward building a proper financial behavior among the people of country after understanding various factors affecting the behavior, as well as positive impacts of good financial behavior which it can bring to the individual as well as economy level [13].

A number of factors play a role in developing financial behavior. Financial literacy is a person's understanding of personal finance issues and is measured by their understanding of various personal finance concepts, like—savings and borrowing concept, insurance, and investment. Financial attitude may be referred to as a state of mind/opinion/judgment for their personal finances, how they treat money and funds, hunger to earn or satisfaction level for their earnings. Locus of control may be referred to as anyone's attitude toward any event that either the same person can control or cannot control the events that occur. Rotter has divided locus of control into internal locus and external locus of control. Internal locus of control is possessing an

Fig. 2 Behaviorism in finance

attitude by an individual assuming his own skills, abilities, and efforts will determine either positive or negative outcomes. While external locus of control is possessing an attitude by an individual that life is majorly dependent on fate, destiny, luck, or other people's support. Risk tolerance is related to investment risk too. People display three types of behavior—high risk (risk seekers), avoid risk (risk averter), and moderate (who are between high risk and avoid risk takers). It is one of the crucial points affecting financial behavior and which somewhere is also because one's knowledge, social behavior, surroundings, standard of living, prospects, etc. Motivation, one's positive or negative motivation toward various financial instruments in the market, can affect in building his/her financial behavior. The people who are positively motivated can easily control spending and purchase impulses and may start/increase in the contribution of savings or investments. Mental accounting, that involves preparing instant rough budget and self-control, also plays a crucial role in affecting financial decisions. Level of income and standard of living, as per recent studies—approx. 46 percent of Indian families have an income of less than ₹15,000 per month [14, 15].

Financial behavior can be improved by following different strategies. Creating a budget using various office software-excel, word, etc., on sheet, on paper, or by various budgeting apps, helps keep track of money received against amount of money spent. Putting aside some amount of money from regular income for important present or future goals helps in managing present funds available along with future funds and needs efficiently. Making necessary payments first is a good way to prioritize the payments for necessities, comfort needs and then luxury needs after saving a good amount for the future. Undertaking credit when only necessary, rather than relying on the present trend of credit being taken at every stage. Investing in the future, for future seen and unforeseen circumstances, prepares individuals for any consequences. Keeping track of credit score (CIBIL) at a fair and better level is also very crucial in today's era. Keeping in mind "if you invest in land, you reap profits for land, if you also invest in yourself, you reap never ending profits" investing in mental and physical health should be the priority. Taking rational decisions which are less or not affected by social pressure prevents financial over-burdening. Invest in place of saving can bring about better returns.

There has always been tension and ambiguity about relation between sociology and psychology given the fact while former focuses on individual perception, cognition, motivation, emotion, and behavior that influence action whereas later is concerned with collectives like families, communities, and so on. Nonetheless, there exists a connection between sociology and psychology as social psychology. The development of social psychology and behavior got a great push with the establishment and functioning of Indian Council of Social Science Research (ICSSR) in 1969 in the form of support received for research, training, and publications in social psychology and behavior [16]. The outcome of these push has led to increase of knowledge related to social behavior like attitude, prejudice, and intergroup relations; social motives and development; social influence processes; poverty, deprivation and social justice; environment and behavior; and social values and development. Besides there has also been many works as opined by resorting to the indigenization-process of making discipline appropriate/suited to the socio-cultural soil of the country, for providing the sizable support to the planning and developmental project to the uplift of the people and progress of the country.

However, what is needed is deeper enquiry theoretically and empirically in order to fathom behavioral dimensions of human social relations and interaction [17].

4 Conclusion

Behavioral science plays a very important role in dealing with every decision in life, whether it is dealing with mental or physical health, financial or economic decisions and other social issues. Behavioral science is the science which deals with applicability of all the inter-related sciences as discussed above. Though much research is going on regarding this, there remain possibilities to enhance knowledge by focusing on various facets in this field, which opens doors to the research gateways.

Acknowledgements We thank friends and colleagues who supported our endeavor to write this paper.

References

1. E.B. Hurlock, *Developmental Psychology* (Tata McGraw Hill Publishing Company, New Delhi, 1978)
2. H.A. Simon, *Models of Bounded Rationality* (MIT Press, Cambridge, MA, 1982)
3. D. Kahneman, Maps of bounded rationality—psychology for behavioural economics. Am. Econ. Rev. **93**, 1449–1475 (2003)
4. A. Tversky, D. Kahneman, The framing of decisions and the psychology of choice. Science **211**(4481), 453–458 (1981)
5. D.G. Goldstein, G. Geiringer, Models of ecological rationality—the recognition heuristic. Psychol. Rev. **109**(1), 75–90 (2002)
6. R.H. Thaler, C. Sunstein, *Nudge—improving decisions about health, wealth, and happiness* (Yale University Press, New Haven, CT, 2008)
7. R.H. Thaler, E.J. Johnson, Gambling with the house money and trying to break even—the effects of prior outcomes on risky choice. Manage. Sci. **36**(6), 643–660 (1990)
8. O. Zellermayer, The pain of paying. Doctoral dissertation, Department of Social and Decision Sciences, Carnegie Mellon University (1996)
9. B. Schwartz, *The Paradox of Choice - Why More is Less* (HarperCollins, New York, 2004)
10. S. Iyengar, M. Lepper, When is choice demotivating—can one desire too much of a good thing? J. Pers. Soc. Psychol. **79**, 995–1006 (2000)
11. D. Ariely, *Predictably Irrational* (HarperCollins, New York, 2008)
12. A. Mulasi, M. Jain, K. Desai, *Predicting the Financial Behavior of Indian Salaried-Class Individuals, Investment Management and Financial Innovations* (2023), pp. 1812–9358. https://doi.org/10.21511/imfi.20(1).2023.03
13. S. Bhatia, S. Singh, Exploring financial well-being of working professionals in the Indian context. J. Financial Serv. Market. (2023). https://doi.org/10.1057/s41264-023-00215-x
14. J. Adriani, Factors affecting financial behaviors—studies in students who do venture creation. Int. J. Rev. Manag. Bus. Entrepreneurship (RMBE) **1**(2), 191–204 (2021). https://doi.org/10.37715/rmbe.v1i2.2427

15. N. Hairin et al., Factors influencing financial management behaviour among university students. Selangor Bus. Rev. **7**(2), 74–87 (2022)
16. Indian Council of Social Science Research https://icssr.org/icssr-glance
17. M. Ameliawati, R. Setiyani, The influence of financial attitude, financial socialization, and financial experience to financial management behavior with financial literacy as the mediation variable, in *International Conference on Economics, Business and Economic Education* (2018), pp. 811–832. https://knepublishing.com/index.php/Kne-Social/article/view/3174

Positive, Negative, or Inconclusive Relationship Between Shareholder's Wealth Creation and Corporate Social Responsibility: A Literature Review

Ruchi Shukla and Ashish Ranjan Sinha

Abstract Extant literature available shows the varied impact of corporate social responsibility (CSR) contribution on corporate financial performance (CFP) thereby its impact on shareholders' wealth creation. This study aims to review past literature to identify the relationship between contribution toward corporate social responsibility and shareholders' wealth creation. For this purpose, research articles which has analyzed the relationship between corporate social responsibility, corporate financial performance, and shareholder wealth creation under have been reviewed. Reasons behind varied results have been analyzed. The study concludes with how future research determining the relationship between CSR and CFP can be shaped to provide conclusive results.

Keywords Corporate social responsibility · Shareholders wealth creation · Negative relation · Positive relation · Corporate financial performance

1 Introduction

Traditional economists have argued that the whole sole responsibility of an organization is to increase its shareholder's wealth [1]. Aligned with this objective most companies, especially those which are in the private sector concentrate to grow their profits. This does not imply that other stakeholders do not play an important role. Businesses are an important part of society and they depend on society to achieve their economic goals [2]. Stakeholder theory which is an important landmark in defining the relationship between the society and the business argues that the success of the business over time will depend upon its good relationship with the stakeholders.

R. Shukla (✉)
National Institute of Technology, Patna, India
e-mail: ruchis.hs19.phd@nitp.ac.in

A. Ranjan Sinha
Department of Humanities and Social Sciences, National Institute of Technology, Patna, India
e-mail: ashish@nitp.ac.in

A. Ismail et al. (eds.), *Advancements in Industry and Technology*,
SpringerBriefs in Applied Sciences and Technology,
https://doi.org/10.1007/978-3-031-97931-6_6

As per general understanding reputation of a company and distinct stakeholder's welfare are important for the wealth maximization and for a company's long-term survival. This paper attempts to highlight the importance of shareholders considering the conventional objective of financial management of a business which is to maximize shareholder's wealth. Although there has been a lot of attempts to study the impact of corporate social responsibility on shareholder's wealth in different type of economies. A meta-analysis of researches which tried to examine the relationship between corporate social responsibility and shareholder wealth maximization has been done and it concluded that a clear relationship has yet not been established due to presence of moderators that may influence the outcomes [3]. The policies and understanding of corporate social responsibility vary from country to country, so the impact of corporate social responsibility on shareholder's wealth will be different from country to country. Let us analyze its impact which also depends upon policies of a nation.

2 Methodology

2.1 Literature Review Inclusion Criterion

We have only included studies that were related to systematically defining relationship between corporate social responsibility and shareholder wealth or corporate social responsibility and corporate financial performance. We included studies done from various regions, and all the studies included were written in English.

2.2 Identification of Literature

We initiated our literature by using the key words "Corporate Social Responsibility" + "Shareholder Wealth" and "Corporate Social Responsibility" + "Corporate Financial Performance." The researcher chose Google Scholar, which is the most frequently used data base for the purpose as the search engine for review of literature related to determination the relationship between CSR and CFP.

Further selection criteria were determined by the title. From the title, if the content seemed fit for the study then that study was chosen for further evaluation. We have chosen only those publications which have been done from 1998 to 2019, as that has been the time period since when CSR has developed into a more prominent concept. Research articles from Scopus indexed journals were considered for the purpose of this study.

A vast previous literature has studied the consequences of spending under CSR on the share prices and hence the value of the firm. Still it is inconclusive. In this study, major findings of various researches done in this context have been analyzed and summed up to conclude that in what conditions the firm's contribution toward CSR has positive impact on shareholder's wealth and vice versa. For the same purpose, depending upon the results of the studies reviewed, they have been grouped into positive, negative, and inconclusive.

Table 1 provides the summary of reasons why CSR has different impact on CFP and shareholders' wealth creation.

Table 1 Analysis of CSR impact on shareholder's wealth

Impact	Reasons
Positive impact	Marketing
	CSR mandatory
	Strategic CSR
	Increase in profitability and performance
	CSR reporting
	Increases trust
Negative impact	Economies where CSR is not mandatory
	Companies which do not spend strategically in CSR
	Investors see CSR spending to result in reduce in stock returns
	Economies where CSR is not mandatory
	Companies which do not spend strategically in CSR
	Investors see CSR spending to result in reduce in stock returns
	In capitalist systems
	No proper communication of CSR spending among stakeholders
	Due to flaws in Indian law (for e.g.: CSR spending is not tax deductible)
	Firms that have age old assets result in having low ratings as regards to CSR
	As organizations have to spend on CSR thus firms which are socially responsible have a competitive disadvantage
Inconclusive impact	Socially responsible firm may get increased profits but simultaneously their expenditures also increases, socially irresponsible firms may see reduced profits but on the other hand their expenditures are also low
	The CFP and CSR relationship is inconclusive because of presence of many moderators which affects the outcome
	Use of corporate resources to create social value which does not affects primary shareholders will not impact the shareholders wealth creation
	Different CSR activities have different impact in shareholder's wealth creation
	Institutional investors do not like spending in securities of socially responsible firms, while investors who are socially conscious may sell off the shares of socially less /irresponsible corporations, thus this buying and selling of shares will result in share price returning to its original
	Age of asset

3 Results and Discussion

3.1 *Positive*

Although recent studies are more inclined toward positive impact, it also depends upon the variable and conditions under which studies were performed.

Let us have a look at some of the studies which concluded with positive impact and how they came to this conclusion.

Firms engaged in environmentally friendly activities and donations have an optimistic impact on shareholders returns. Market interprets various social actions in different ways. Social actions have been divided into four categories donations, recycling, social policy, and environmentally friendly activities. If the actions of a firm are socially responsible, then it can have a constructive impact on the value of a firm [4]. The importance of strategic CSR activities has also been highlighted and to implement the same the top management has to play an active role. This study also proposes that ethical CSR which revolves around ethical duties and responsibilities is mandatory [5]. The analysis of the relationship between CSR and other financial parameters especially at the time of global financial crisis concluded that the firms which are high on social capital, earn greater trust from stakeholders especially in periods when the overall trust on corporations and market suffers a negative shock [6]. Along with improving firm's reputation and social value, CSR also improves performance and profitability. In India, companies which are socially responsible perform better. And not only contributing to CSR but also properly reporting is important. Companies can develop competitive advantages by strategically practicing CSR. A research which took sample of listed companies in Bucharest Stock Exchange in the time period between 2008 and 2011 and it was done to find out the relationship between CSR and firm performance. It was concluded that although there is a negative relationship between CSR and return on sales (ROS) there is a positive relationship between CSR and EPS (earning per share) [7].

A research was done to understand the relationship between the performance, charitable contributions, and market value of listed companies in Romania. The results of the study concluded that there exists a positive relationship between the firm's performance, its charitable contributions and the market value of the Romanian listed companies [8]. A research was done to study whether awareness of CSR which can be measured by media coverage, moderates the relationship between financial and social performance. The results support the notion that CSR awareness among the stakeholders affects the manner in which CSR awareness can result in financial gains [9].

It highlights the importance of two things:

(a) Firm's contribution to social initiatives.
(b) How important it is to communicate about the CSR initiatives to primary stakeholders.

In a study which was done to determine the role of marketing on CSR-shareholder wealth creation, it was concluded that marketing plays an important and complementary role in the CSR-shareholder relationship. The results provide enough evidence that CSR has less potential to effect firm shareholder wealth directly but can do so with right marketing efforts [10]. Shareholders and brokers both of them agree to the point that the companies which are involving in high level of CSR activities enjoy a higher level of confidence of investors which can be seen through the increase in the stock prices of such companies and ultimately it increases the goodwill [11]. Legal CSR increases firm value while normative CSR decreases firm value [12]. The firms that are more responsible toward the environment face an increase in their stock prices, whereas the firms that are irresponsible face notable decrease in their stock prices. A study which was done to analyze the dividend payment by firms concluded that firms which are high on CSR pay more dividends than firms which are low on CSR. The payout of dividends is more stable in case of firms that are actively engaged in high CSR activities, and as a result, socially responsible firms do not have to frequently adjust their dividends, whereas socially irresponsible firms do have to [13].

3.2 Negative Impact

Contributing toward CSR has been made mandatory by India and it is the first country in the world which has implemented CSR as a law. Section 135 of Companies Act 2013 lays down that if the net worth of a company is more than Rs.500 Crore, or the net profit is more than Rs.5 Crore or the turnover is more than Rs.1000 crore then it has to mandatorily spend 2% of average net profits of past three years toward CSR activities. From accountability point of view, if we study the impact of CSR on investors then, it turns out to be a loss for them, because CSR is allowed to be charged as an expense in the profit and loss account, which in turn reduces the net profit and hence the dividend to be distributed to the shareholders. The shareholders who are there to earn constant income, in turn, start selling off their share which results in a reduction in prices of shares. The countries where CSR has been made legal, even if the companies are bearing losses in the current financial year, but fulfils any of the criteria to become legible to contribute in CSR activities, has to do so, which puts an extra financial burden on the organizations [14]. The Italian investors perceive the contribution toward corporate social performance as avoidable expenses and consider it to reduce the shareholder's income and companies value which thereby reduces the stock price and hence conclude it as a negative relationship between stock prices and corporate social performance. The reason behind such behavior can be a capitalist system, their cultural background and CSR not legalized [15]. The impact of CSR on shareholder's wealth also depends upon the law of the land. Socially responsible firms do not have competitive advantage, which is because of the increase in their expenses which they incur for contributing in CSR.

The studies that have been analyzed for the purpose of this research from different angles resulted in stating that either the relationship is mixed, i.e., both negative and positive or there is an insignificant relationship. Thus, there lacks consistency in the results which gives further scope to researchers to tackle this issue from other different perspectives. By analyzing specially, the recent studies where results have been concluded to be positive, it can be said that since CSR in many countries have been legalized and also it has been made mandatory to do proper CSR reporting, thus it has resulted in positive impact of CSR spending on the shareholder's wealth. The countries where CSR is not strategically done, there shareholders do not come to know the benefits of such expenditure and hence see it from the accountability point of view, and from that view they see decrease in profitability, and also decrease in earnings per share of the shareholders. Some studies have shown that there cannot be a common conclusion when linking CSR with financial performance, thus now the finance researchers have started taking various aspects of financial performance and then have started studying the relationship between CSR and that particular aspect in detail. When different aspects of financial performance are linked with CSR, then results of such studies are concluded to be positive, negative as well as not related to CSR [16]. The previous literature review suggests that the studies done in this area have been broadly from two main aspects. The first approach is from the point of view of market valuations and argues that CSR impacts the market value of the firm through risk and expected return, future income profitability. Thus by studying the specific channels, a researcher can come to the conclusion that how CSR increases or worsens different parameters of corporate financial performance. CSR and various financial aspects such as investment, dividend decisions [17], corporate cash holdings and capital structure mergers and acquisitions should be researched separately for a definite conclusion defining their relationship. Thus, the finance scholars and researchers in order to arrive at more robust results have to study the impact of CSR on various financial parameters and not only this the result have to be concluded based upon economy to economy, depending upon whether CSR policy is normative or legalized in a particular economy.

4 Conclusion

The studies that have been analyzed for the purpose of this research from different angles resulted in stating that either the relationship is mixed, i.e., both negative and positive or there is an insignificant relationship. Thus, there lacks consistency in the results which gives further scope to researchers to tackle this issue from other different perspectives. By analyzing specially, the recent studies where results have been concluded to be positive, it can be said that since CSR in many countries have been legalized and also it has been made mandatory to do proper CSR reporting, thus it has resulted in positive impact of CSR spending on the shareholder's wealth. The countries where CSR is not strategically done, there shareholders do not come to know the benefits of such expenditure and hence see it from the accountability point of view, and from that view they see decrease in profitability, and also decrease in earnings per share of the shareholders. Some studies have shown that there cannot be a common conclusion when linking CSR with financial performance, thus now the finance researchers have started taking various aspects of financial performance and then have started studying the relationship between CSR and that particular aspect in detail. When different aspects of financial performance are linked with CSR, then results of such studies are concluded to be positive, negative as well as not related to CSR [16]. The previous literature review suggests that the studies done in this area have been broadly from two main aspects. The first approach is from the point of view of market valuations and argues that CSR impacts the market value of the firm through risk and expected return, future income profitability. Thus by studying the specific channels, a researcher can come to the conclusion that how CSR increases or worsens different parameters of corporate financial performance. CSR and various financial aspects such as investment, dividend decisions [17] corporate cash holdings [18] and capital structure mergers and acquisitions should be researched separately for a definite conclusion defining their relationship. Thus, the finance scholars and researchers in order to arrive at more robust results have to study the impact of CSR on various financial parameters and not only this the result have to be concluded based upon economy to economy, depending upon whether CSR policy is normative or legalized in a particular economy.

References

1. M. Friedman, A Friedman doctrine: the social responsibility of business is to increase its profits. The New York Times Mag. **13**(1970), 32–33 (1970)
2. A. Crane, A. McWilliams, D. Matten, J. Moon, D.S. Siegel, *The Oxford Handbook of Corporate Social Responsibility*. (Oxford University Press on Demand, 2008)
3. M. Orlitzky, F.L. Schmidt, S.L. Rynes, Corporate social and financial performance: a meta-analysis. Organ. Stud. **24**(3), 403–441 (2003)
4. P.L. Hall, R. Rieck, The effect of positive corporate social actions on shareholder wealth. J. Financ. Strateg. Decis. **11**(2), 83–89 (1998)
5. G.P. Lantos, The boundaries of strategic corporate social responsibility. J. Consumer Market. (2001)
6. K.V. Lins, H. Servaes, A. Tamayo, Social capital, trust, and firm performance: the value of corporate social responsibility during the financial crisis. J. Financ. **72**(4), 1785–1824 (2017)
7. L.N. Simionescu, Ă.C. Gherghina, Corporate social responsibility and corporate performance: empirical evidence from a panel of the Bucharest Stock Exchange listed companies. Manag. Market. **9**(4) (2014)
8. C.D. Hategan, R.I. Curea-Pitorac, Testing the correlations between corporate giving, performance and company value. Sustainability **9**(7), 1210 (2017)
9. Y. Rhou, M. Singal, Y. Koh, CSR and financial performance: the role of CSR awareness in the restaurant industry. Int. J. Hosp. Manag. **57**, 30–39 (2016)
10. S. Mishra, S.B. Modi, Corporate social responsibility and shareholder wealth: tThe role of marketing capability. J. Mark. **80**(1), 26–46 (2016)
11. M. Kansal, M. Joshi, Perceptions of investors and stockbrokers on corporate social responsibility: a stakeholder perspective from India. Knowl. Process. Manag. **21**(3), 167–176 (2014)

12. M.A. Harjoto, H. Jo, Legal vs. normative CSR: differential impact on analyst dispersion, stock return volatility, cost of capital, and firm value. J. Bus. Ethics **128**, 1–20 (2015)
13. M. Benlemlih, Corporate social responsibility and dividend policy. Res. Int. Bus. Financ. **47**, 114–138 (2019)
14. G.K. Kapoor, S. Dhamija, Mandatory CSR spending—Indian experience. Emerg. Econ. Stud. **3**(1), 98–112 (2017)
15. M.F. Izzo, F. di Donato, The relation between corporate social responsibility and stock prices: an analysis of the Italian listed companies. Available at SSRN 1986324 (2012)
16. M.E. Porter, M.R. Kramer, The link between competitive advantage and corporate social responsibility. Harv. Bus. Rev. **84**(12), 78–92 (2006)
17. M.T. Rakotomavo, Corporate investment in social responsibility versus dividends? Soc. Respons. J. **8**(2), 199–207 (2012)
18. M. Arouri, G. Pijourlet, CSR performance and the value of cash holdings: international evidence. J. Bus. Ethics **140**, 263–284 (2017)

Optimizing Investment Portfolio Selection with the Hybrid Conjugate Gradient Method

Melly Camelia Mohadi and Nur Syarafina Mohamed

Abstract The conjugate gradient (CG) method is an optimization method which is commonly applied for solving unconstrained optimization problems due to its simplicity, properties of simplicity for low memory requirements and it is Hessian free. The CG method has been developed and improved in terms of efficiency by undergoing numerous modifications from time to time. An efficient investment portfolio is important for all investors to keep track of the best investment following the current situation. A hybrid CG method is used in optimizing the investment portfolio selection. Numerical comparison associated with unconstrained test functions as benchmark between six hybrid CG methods is done to investigate the efficiency of the methods by using exact line search. As a result, the hybrid-Syarafina-Mustafa-Rivaie (HSMR) method is considered as the most efficient method. Thus, the HSMR method is being applied as optimization method in obtaining the optimal investment portfolio selection.

Keywords Hybrid conjugate gradient method · Exact line search · Unconstrained optimization · Investment portfolio selection

1 Introduction

An optimization method can be defined as an alternative or technique to seek the best values or solutions over a range of inputs with the aim of maximizing or minimizing an objective function [1]. Generally, optimization problems can be classified into two group-unconstrained and constrained optimization problems. The process of

M. C. Mohadi · N. S. Mohamed (✉)
Department of Mathematical Sciences, Faculty of Science, Universiti Teknologi Malaysia, Johor Bahru, Johor, Malaysia
e-mail: nursyarafina@utm.my

M. C. Mohadi
e-mail: mellycamelia@graduate.utm.my

A. Ismail et al. (eds.), *Advancements in Industry and Technology*,
SpringerBriefs in Applied Sciences and Technology,
https://doi.org/10.1007/978-3-031-97931-6_7

maximizing the objective function with regard to several variables while taking the variable restrictions into account is known as constrained optimization.

Apposed from that, unconstrained optimization simply indicates that the variable's choice can take on any value without any restrictions [2].

The stock market is considered as one type of the financial market as it is produced through the purchase and sale of various financial instruments. An efficient investment portfolio is important for all investors to keep track of the best investment following the current situation. To get the best portfolio, portfolio optimization is needed, where it optimizes the portfolio by minimizing the financial risk [3], by using the hybrid conjugate gradient (CG) method as the optimization method. The CG method is an iterative method that provides the optimal solution when there is no rounding-off error [4]. A quadratically convergent gradient approach was described for identifying the local minimum of a function of several variables [5]. Consequently, a nonlinear CG approach was presented. CG techniques use up less storage capacity, simpler implementation and are Hessian free [6].

The purpose of this study is to investigate the efficiency of hybrid CG methods by doing numerical comparison on six existing hybrid CG methods. Also, this study aimed to use the hybrid CG method that yields the most efficient numerical results into the application, which is optimizing the investment portfolio selection.

2 Methodology

Combining at least two established traditional CG techniques results in a hybrid CG approach [7]. The first hybrid CG method was proposed, where this hybrid CG method is a combination of the Fletcher-Reeves (FR) method and the Polak-Ribiere-Polyak (PRP) method [8]. An alternative was made by generating a new hybrid CG method with the combination of the same methods, PRP and FR method. It focused on allowing the coefficients to be negative, instead of being restricted in being non-negative [8]. There are more well-known hybrid CG methods, such as the hybrid Dai-Yuan (HDY) method. The HDY method is a modification method that combines its algorithm with the Hestenes-Stiefel (HS) method [9]. The approach demonstrated effective numerical performance and global convergence under Wolfe line search [10]. Both Wei-Yao-Liu (WYL) and Polak-Ribiere-Polyak (PRP) methods come with restart properties and satisfied the descent condition under strong Wolfe search. Inspired by these factors, a new hybrid CG method was proposed, known as the Yasir-Hamoda-Mamat (YHM) method, a combination of WYL and PRP methods [11]. Another hybrid CG method also recently proposed, a combination of Syarafina-Mustafa-Rivaie (SMR) and Rivaie-Mustafa-Ismail-Leong (RMIL) methods.

Known as the hybrid-Syarafina-Mustafa-Rivaie (HSMR) method, the idea was initiated by the factor of restart property. If the value tends to be negative, it is recommended to coordinate the beta coefficient as 0 [12]. This proposed method (1)–(6) satisfies the sufficient descent property under inexact strong Wolfe line search [6].

$$\beta_k^{\text{hDY}} = \max\left\{0, \min\left\{\beta_k^{\text{HS}}, \beta_k^{\text{DY}}\right\}\right\} \tag{1}$$

$$\beta_k^{\text{HJHJ}} = \frac{g_k^2 - \max\left\{0, \frac{g_k}{g_{k-1}}\right\} g_k^T g_{k-1}}{\max\left\{g_{k-1}^2, d_{k-1}^T (g_k - g_{k-1})\right\}} \tag{2}$$

$$\beta_k^{\text{YHM}} = \begin{cases} \beta_k^{\text{PRP}} & \text{if } 0 \le g_k^T g_{k-1} \le \left\|g_k^2\right\| \\ \beta_k^{\text{WYL}} & \text{otherwise} \end{cases} \tag{3}$$

$$\beta_k^{\text{HSMR}} = \max\left\{0, \min\left\{\beta_k^{\text{SMR}}, \beta_k^{\text{RMIL}}\right\}\right\} \tag{4}$$

$$\beta_k^{\text{HSNHMR}} = \max\left\{0, \ \min\left\{\beta_k^{\text{HS}}, \beta_k^{\text{NHMR}}\right\}\right\} \tag{5}$$

$$\beta_k^{\text{HSLAMR}} = \begin{cases} \beta_k^{\text{HS}} & \text{if } 0 \le \beta_k^{\text{LAMR}} \le \beta_k^{\text{HS}} \\ \beta_k^{\text{LAMR}} & \text{otherwise} \end{cases} \tag{6}$$

2.1 Unconstrained Optimization Problem

The hybrid CG method was chosen as the optimization method in solving unconstrained optimization problem as below.

$$\min_{x \in R^n} f(x)$$

where $f : \mathbb{R}^n \to \mathbb{R}$ is continuously differentiable, $\mathbb{R}^n$ indicates that it is the n-dimensional Euclidean space, and the gradient is defined by $g(x) = \forall f(x)$.

2.2 Iteration

The CG method is an iterative technique that yields approximation points for the following formula at each step:

$$x_{k+1} = x_k + \alpha_k d_k; \, k = 0, 1, 2, \ldots \tag{7}$$

where x_k and x_{k+1} are previous and current points, respectively.

2.3 Search Direction

The search direction, d_k is defined as below.

$$d_k = \begin{cases} -g_k & \text{if } k = 0 \\ -g_k + \beta_k d_{k-1} & \text{if } k \geq 1 \end{cases} \tag{8}$$

for a standard CG method.

2.4 Exact Line Search

In this study, exact line search will be used as shown in Eq. (9) below.

$$f(x_k + \alpha_k d_k) = \min_{\alpha \geq 0}(x_k + \alpha d_k) \tag{9}$$

where the above line search is evaluated by taking the derivative with respect to α and the equation is set to 0 to solve for α. d_k is the search direction while x_k is the current initial point used.

2.5 Algorithm of CG Method

Below are the six crucial steps in the computational process [6]:

Step 1: Initialization. Start with $k = 0$, which generates x_o.

Step 2: Solutions according to Eqs. (1), (2), (3), (4), (5), (6) are computed.

Step 3: Search direction, which is based on (8) is computed. Stop the process when $g_k = 0$.

Step 4: Step size is also computed, based on Eq. (9).

Step 5: New point is generated as in Eq. (7).

Step 6: Convergence and stopping criteria is checked. Stop when $f(x_{k+1}) < f(x_k)$ and $g_k \leq \varepsilon$, otherwise return to step 1 by $k = k + 1$.

3 Results and Discussion

Numerical comparison between six existing hybrid CG methods as shown from Eqs. (1)–(6) are conducted under exact line search associated with 13 unconstrained test functions as benchmark. A total of four different initial points with a range of dimensions are chosen for the test functions as shown in Table 1 which is being referred by [6, 13, 14].

Table 1 List of test functions with dimensions and initial points

No.	Test function	Dimension	Initial points
1	Booth	2	(3,3), (6,6), (9,9,), (12,12)
2	Treccani	2	(5,5), (10,10), (12,12), (20,20)
3	Leon	2	(2,2), (5,5), (8,8), (10,10)
4	Diagonal 4	2	(5,5), (7,7), (20,20), (45,45)
5	Dixon and price	2,4	(2,…,2), (10,…,10), (14,…,14), (22,…,22)
6	Shallow	2,4,10	(2,…,2), (5,…,5), (12,…,12), (20,…,20)
7	Fletchcr	2,4,10	(5,…,5), (6,…,6), (10,…,10), (20,…,20)
8	Sum square	2,4,10	(2,…,2), (4,…,4), (10,…,10), (16,…,16)
9	Extended Rosenbrock	2,4,10	(4,…,4), (11,…,11), (14,…,14), (30,…,30)
10	Extended Beale	2,4,10,100	(3,…,3), (4,…,4), (5,…,5), (8,…,8)
11	Extended Denschnb	2,4,10,100	(5,…,5), (10,…,10), (25,…,25), (200,…,200)
12	Extended White and Hoist	2,4,10,500,1000	(2,…,2), (5,…,5), (8,…,8), (9,…,9)
13	Extended Himmelblau	2,4,10,500,1000	(3,…,3), (6,…,6), (14,…,14), (22,…,22)

The number of iteration (NOI) and central processing unit (CPU) time for each sample are collected. The results are compared based on the performance profile proposed by Dolan and Moore [15]. The more test problems solved indicates that the method is said to be robust [6]. Figure 1 shows that the HSMR method has the best efficiency in terms of solving the NOI as at the top left graph. Despite of being lower than some methods at the right side of the graph, the HSMR method showcases its ability in solving all the test functions. Based on Fig. 2, the HSMR method shows the best efficiency in terms of solving the test functions as it is placed at the top right graph, with the shortest CPU time as at the top left graph.

3.1 Application

A total of seven top stocks listed in Bursa Malaysia are selected as an investment portfolio. For this study, the daily closing price of the stock is used with a duration of 2 years (Jan 1, 2020—Dec 31, 2022) [16].

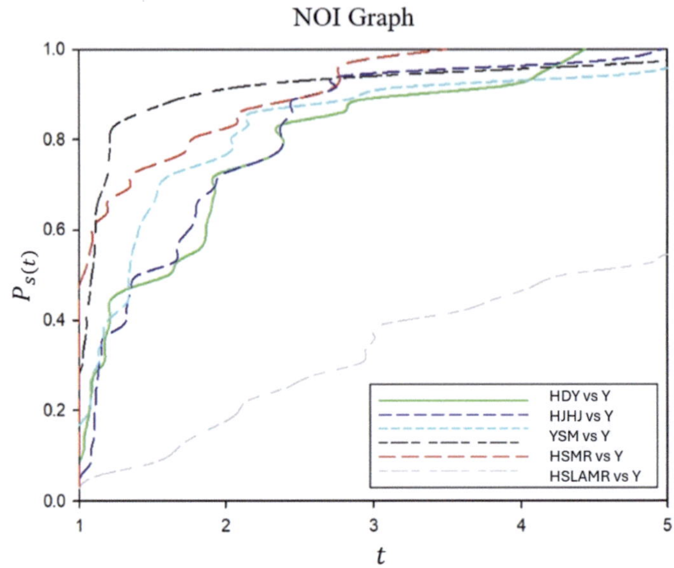

Fig. 1 Number of iteration (NOI) graph

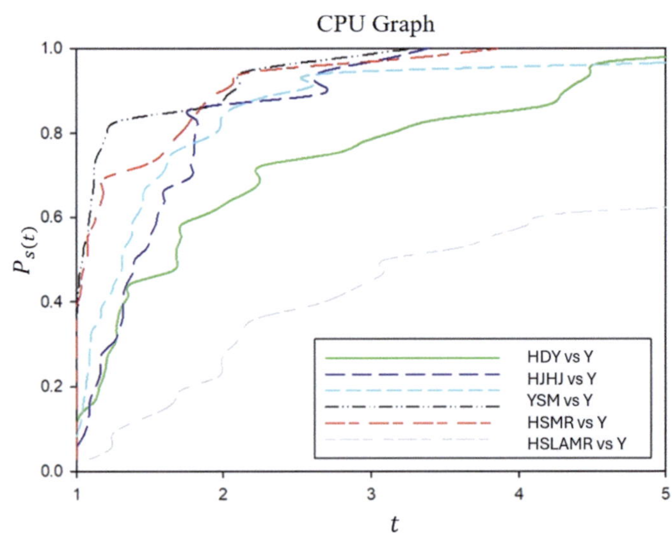

Fig. 2 Central processing unit (CPU) time graph

Based on the optimization model, the Eq. (10) is transformed into an unconstrained optimization problem by using manipulation algebra. By letting $w_7 = 1 - w_1 - w_2 - w_3 - w_4 - w_5 - w_6$, the unconstrained optimization problem can be obtained as below.

$$\begin{aligned}
\min_{(w_1, w_2, w_3, w_4, w_5, w_6) \in \mathbb{R}^6} \quad & (w_1 + w_2 + w_3 + w_4 + w_5 + w_6 - 1)(0.000138w_1 \\
& +0.000112w_2 + 0.000126w_3 + 0.000146w_4 + 0.000109w_5 + 0.000121w_6 \\
& -0.000185) + w_1(0.005557w_1 + 0.000097w_2 + 0.000045w_3 + 0.000039w_4 \\
& -0.000001w_5 + 0.000027w_6 + 0.000047) + w_2(0.001697w_1 + 0.000429w_2 \\
& +0.000111w_3 + 0.000077w_4 + 0.000032w_5 + 0.000062w_6 + 0.000074) \\
& +w_3(0.000032w_1 + 0.000126w_2 + 0.000391w_3 \\
& +0.000047w_4 + 0.000046w_5 + 0.000037w_6 + 0.000059) \\
& +w_4(0.000046w_1 + 0.000111w_2 + 0.000067w_3 + 0.000097w_4 \\
& +0.000006w_5 + 0.000037w_6 + 0.000040) \\
& +w_5(0.000029w_2 - 0.000031w_1 + 0.000028w_3 - 0.000031w_4 \\
& +0.000181w_5 - 0.000022w_6 + 0.000077) \\
& +w_6(0.000010w_1 + 0.000071w_2 + 0.000031w_3 \\
& +0.000012w_4 - 0.000011w_5 + 0.000127w_6 + 0.000065)
\end{aligned} \tag{10}$$

From Table 2, the proportion value for HLB is 0.1900. This indicates that 19% of total allocation of funds are formed by HLB. The second and third proportion have negative value of proportion, which are − 0.1198 and − 0.0039, respectively. Negative sign simply presents that the investor is short selling. This indicates that the amount of funds from GEN and PTG are − 11.98% and − 0.39%, respectively. MBB has the proportion value of 0.3915. The amount of funds that will be allocated is 39.15%. The proportion value for KKK is 0.1578, this means that a total of 15.78% of funds will be allocated. The sixth proportion for TNB is 0.1546, at which the funds allocated are 15.46%. The last proportion for PPB is 0.2268, which allocates 22.68% of the total allocation of funds. By allocating the value proportion of each stock from Table 2, the investment portfolio provides a rate of return of 0.0044% for the total allocated funds and the risk of return of 0.008% on total funds.

Table 2 Optimal portfolio weight composition of each stock

Stock	HLB	GEN	PTG	MBB	KKK	TNB	PPB
Proportion	0.1900	− 0.1198	− 0.0039	0.3915	0.1578	0.1546	0.2268

4 Conclusion

In conclusion, the HSMR method exhibits the most efficient numerical results in terms of NOI and CPU time from numerical comparison between hybrid CG methods. Furthermore, optimal weight composition for each stock is obtained by using the HSMR method in solving the risk function. Thus, optimal investment portfolio is achieved.

Acknowledgements This work is supported by UTM Fundamental Research (UTMFR) grant with a reference of (PY/2022/02418, Q.J130000.3854.22H10).

References

1. D.T.M.Salih, B.M. Faraj, Comparison between steepest descent method and conjugate gradient method by using Matlab. J. Stud. Sci. Eng. **1**(1), 20–31 (2021). https://doi.org/10.53898/josse2021113
2. G. Cornuejols, R. Tütüncü, *Optimization Methods in Finance*. Cambridge University Press (2006)
3. J.C. Francis, D. Kim, *Modern Portfolio Theory : Foundations, Analysis, and New Developments + Website*. Wiley (2013). https://ci.nii.ac.jp/ncid/BB17944884
4. M.R. Hestenes, E. Stiefel, Methods of conjugate gradients for solving linear systems. J. Res. Natl. Bur. Stand. **49**(6), 409 (1952). https://doi.org/10.6028/jres.049.044
5. R.R.M. Fletcher, C. Reeves, Function minimization by conjugate gradients. Comput. J. **7**(2), 149–154 (1964). https://doi.org/10.1093/comjnl/7.2.149
6. N. Zullpakkal, N. Aini, N.H.A. Ghani, N.S. Mohamed, N. Idalisa, M. Rivaie, Covid-19 data modelling using hybrid conjugate gradient method. J. Inf. Optimization Sci. **43**(4), 837–853 (2022). https://doi.org/10.1080/02522667.2022.2060610
7. S. Devila, M. Malik, W. Giyarti, A new hybrid Prp-Mmsis conjugate gradient method and its application in Portofolio selection. Jurnal Riset Dan Aplikasi Matematika (JRAM) 5(1), 47 (2021). https://doi.org/10.26740/jram.v5n1.p47-59
8. C. Storey, D. Touati-Ahmed, Efficient hybrid conjugate gradient techniques. J. Optim. Theory Appl. **64**(2), 379–397 (1990). https://doi.org/10.1007/bf00939455
9. M. Malik, M. Mamat, S.S. Abas, M.I. Ibrahim, Solving unconstrained minimization problems with a new hybrid conjugate gradient method, in *Proceedings of the 5th NA International Conference on Industrial Engineering and Operations Management Detroit* (Michigan, USA, 2020)
10. J. Jian, X. Jiang, H. Lin, A hybrid conjugates gradient method with descent property for unconstrained optimization. Appl. Math. Model. **39**(3–4), 1281–1290 (2015). https://doi.org/10.1016/j.apm.2014.08.008
11. Y. Salih, M.A. Hamoda, M. Rivaie, New hybrid conjugate gradient method with global convergence properties for unconstrained optimization. Int. J. Eng. Technol. **7**(3.28), 54–57 (2018)
12. N.S. Mohamed, M. Mamat, M. Rivaie, S.M. Shaharudin, A comparison on classical hybrid conjugate gradient method under exact line search. Int. J. Adv. Intell. Inf. **5**(2), 150 (2019). https://doi.org/10.26555/ijain.v5i2.356
13. N. Zull, N. Aini, S. Shoid, N.H.A. Ghani, N.S. Mohamed, M. Rivaie, M.H. Mamat, A conjugate gradient method with descent properties under strong Wolfe line search. J. Phys. **890**, 012105 (2017). https://doi.org/10.1088/1742-6596/890/1/012105

14. M. Jamil, X. Yang, A literature survey of benchmark functions for global optimization problems. Int. J. Math. Modelling Numerical Optim. **4**(2), 150 (2013). https://doi.org/10.1504/ijmmno.2013.055204
15. E.M. Dolan, J.J. Moré, Benchmarking optimization software with performance profiles. Math. Program. **91**(2), 201–213 (2002). https://doi.org/10.1007/s101070100263
16. Yahoo Finance. Yahoo Finance—Business Finance, Stock Market, Quotes, News. Yahoo Finance. https://finance.yahoo.com/ (2022). Accessed Nov 2023

Investigation of Leakage Inside an Underground Pipeline Using the SMART Nuclear Technique

Noraishah Othman, Ahmad Zubair Johan, Muhammad Anas Mat Marzuki, W. Mohd Wafiy W. Zakaria, and Muhamad Norfazli Muhamad Amin

Abstract A two-module system that incorporates a radiation detector and tracking system has been developed to localize the leakage in underground industrial pipelines using the nuclear technique. This is the state of the art of radiotracer approach which overrules the conventional radiotracer technology in on-line determination of leak location. A conventional technology is cumbersome and laborious, time consuming and expensive. That implies the number of dug pits that should be excavated to assign an individual detection system, number of detectors to be assigned with respect to the length of buried pipe and the number of manpower to be deployed to supervise and monitor the radiation works. Meanwhile, the SMART nuclear technique is cost effective, requires minimal supervision, provides fast results and is accurate. The water proof of a 2-module encapsulated system is launched at the upstream of the pipeline and picks up the radiation signals that mimics the leak points and the data is stored in the data logger. Upon retrieval of the system, the data is analyzed to determine the exact location of the stealth leak. In this study, a dry experiment was conducted to assess the performance of the developed prototype.

Keywords Two module system · Radiotracer · Leak detection · SMART nuclear technique · Underground pipeline

N. Othman (✉) · A. Z. Johan · M. A. Mat Marzuki
Plant Assessment Technology, Industrial Technology Division, Kajang, Selangor, Malaysia
e-mail: noraishah@nm.gov.my

A. Z. Johan
e-mail: zubair@nm.gov.my

M. A. Mat Marzuki
e-mail: anas@nm.gov.my

W. M. W. W. Zakaria · M. N. Muhamad Amin
IOI CONEZION Connection Commercial, Putrajaya, Sepang, Selangor, Malaysia
e-mail: wafiy@exisgroup.com.my

M. N. Muhamad Amin
e-mail: mnfazli.amin@raudatech.com

A. Ismail et al. (eds.), *Advancements in Industry and Technology*,
SpringerBriefs in Applied Sciences and Technology,
https://doi.org/10.1007/978-3-031-97931-6_8

1 Introduction

Leak detection using radiotracer techniques is one of the most widespread applications of nuclear technique in industry. Nuclear Malaysia is the local industrial radiotracer service provider that owns the reactor TRIGA PUSPATI (RTP) in Malaysia. The detection of leakage in underground pipelines is a significant challenge to many industries, in particular to the oil and gas industry. Therefore, Nuclear Malaysia has adopted the radiotracer technology (RT) to locate and verify the leakage location. RT is the introduction of radiotracer inside the domain and the tracing of radiation is conducted externally using a scintillation detector. Any drop of velocity from upstream to downstream fluid flow is susceptible to potential leak. RT is very sensitive, effective, and competitive for on-line leak detection [1, 2]. However, the conventional radiotracer technique is very expensive, requires a lot of manpower, cumbersome and very time consuming [3]. Thus, a two-module prototype is introduced to address the aforementioned problems. It is the state of the art of RT and is developed to detect underground pipeline leakage. It consists of a scintillation detector, data logger, and battery to power up the system and is encapsulated in waterproof capsule. Odometers are mounted onto the capsule body externally. Programming is done to integrate the odowheel and detector for data acquisition and storage in the data logger. The radioactive signals are captured by detectors whereas the distance traveled by capsule is provided by odowheel.

The leak detection is obtained by observing the presence of a peak. There are two type of leak detection systems: Static and dynamic detection [4]. Static leak detection is the case where dedicated sensors are installed in the vicinity of the piping area and valves which are capable to intermittently acquire data and transmit the information to the headquarters whereas dynamic leak detection is deploying a mobile system inside the suspected leak pipeline. Thus, the developed prototype is categorized under the dynamic leak detection system. The sensing mechanism utilizes a silicon photomultipliers (SiPM) sensor that produces feedback output in the presence of radioactive material. The system is launched at the upstream of the pipeline and picks the signals and stores them inside the data logger. Upon retrieval of the system, the analysis is carried out to identify the location of the respective radioactive material which implies the location of leak at designated speed. This prototype consists of two modules as shown in Fig. 1. Module 1 is the radioactive tracer module whereas module 2 is the tracking module, respectively.

2 Methodology

In this study, the prototype is tested under lab condition that means pull through test (PTT) is adopted under dry condition (i.e., no fluid domain involves in the piping system). The modules with a mass of approximately 60 kg is pulled by winch system with capacity of 1814 kg as shown in Fig. 2. The prototype is tested on two different

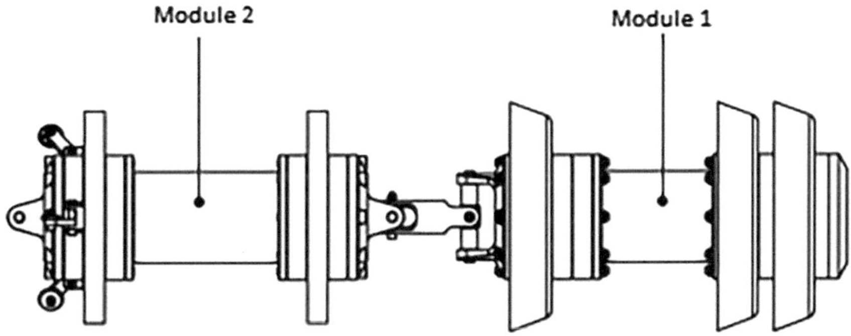

Fig. 1 Prototype of SMART nuclear system

skids. Skid one consists of 25.4 cm inner diameter with two identical 4.5 m straight pipes attached together whereas skid two is a S-bend pipe with 1 m straight pipe attach to its end as shown in Fig. 3.

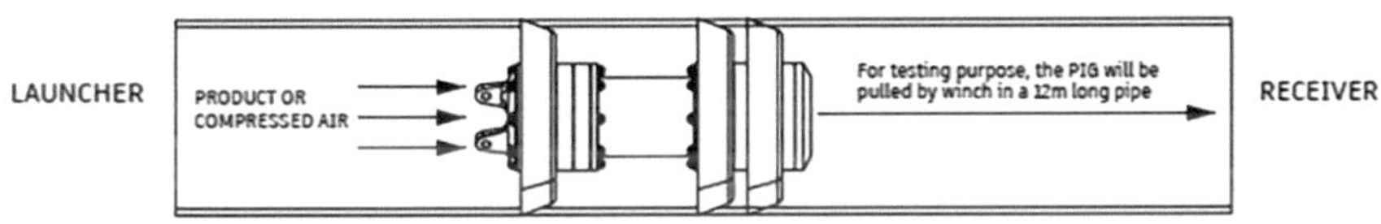

Fig. 2 Pull through test (PTT) mechanism

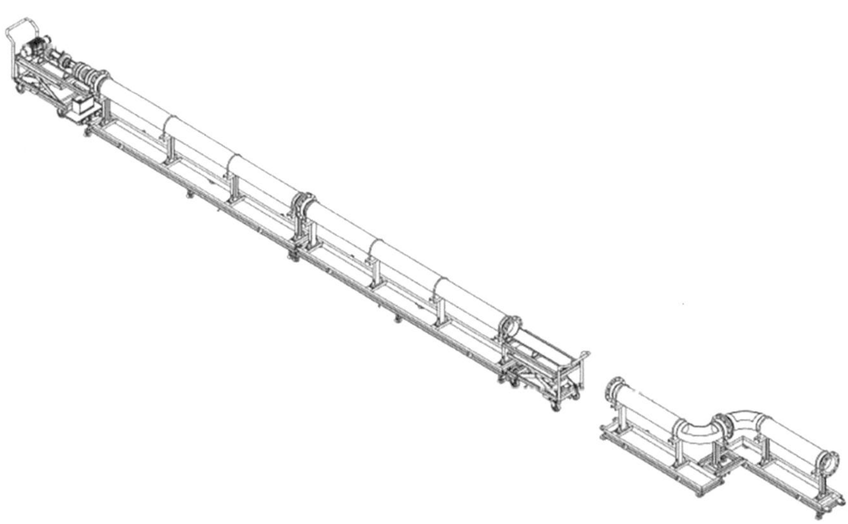

Fig. 3 Schematic diagram of skid 1 and skid 2, respectively

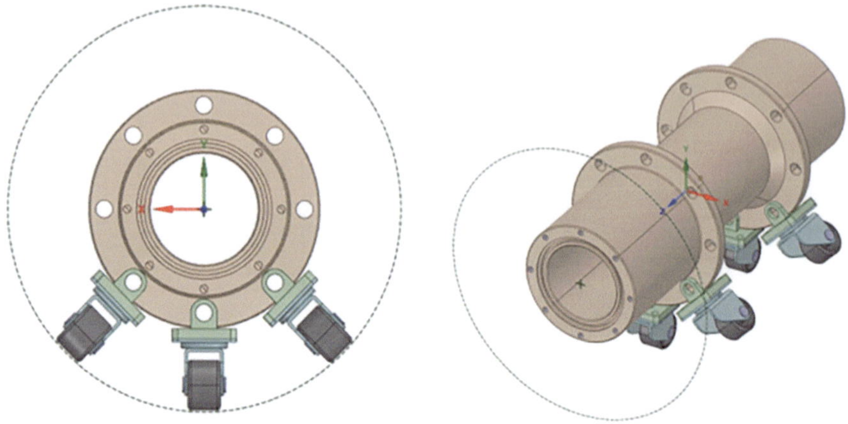

Fig. 4 Example of installation of wheels

2.1 *Acquisition of Radiation Signals*

Each skid has its own objective in which skid 1 is meant to test the ability of the prototype to acquire radioactive signals (Cs-137) at fixed speed. Cs-137 is a by-product of the fission process in a nuclear reactor and decays via beta (0.514 MeV) and gamma emission (0.662 MeV) with a half-life of 30.07 years [1, 2]. Each individual Cs-137 is securely shield and mounted on the pipe wall externally at designated distant while the prototype is pulled inside the pipe using a winch. Various type of Cs-137 activities are used in this study which is divided into two categories (i.e., high and low activities). In this study, activity is the concentration or the strength of the emitted gamma radiation. The radiation is captured by the SiPM detector and converted into numerical value (counts). Table 1 shows the categorization of activities and parameters involved and speed is measured by the distance travel by the system over time. In order to reduce severe abrasion and friction as well as damage of the polyurethane rubber during multiple testing at skid 1, the polyurethane (PU) rubber cup is removed. Instead, wheels are used as substitute to ease the mobility of the prototype inside the pipe as shown in Fig. 4.

2.2 *Maneuverability of Two-Module System*

Skid 2 is used to check the maneuverability of the two-module system inside the bend pipe. The S-bend pipe is used to verify the capability of the prototype to bend along the 90° pipe since it is designed with the most crucial bend tolerance which is 1.5D pipe as shown in Fig. 5. A 1.5D pipe bend is a pipe bend with a bending

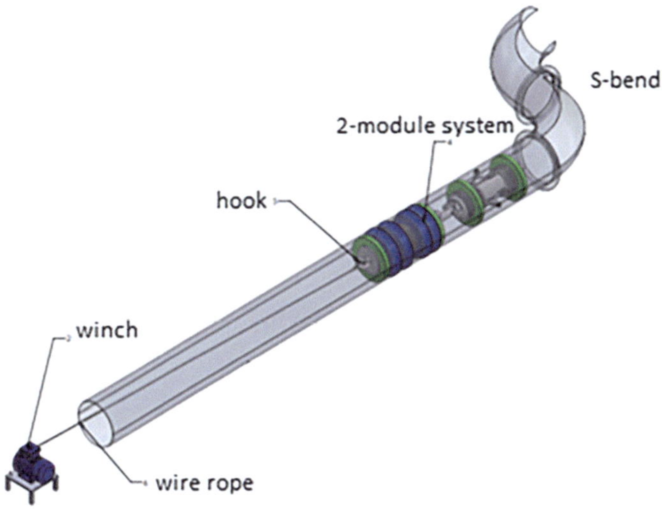

Fig. 5 Mobility of two-module system inside 1.5D bend pipe

radius equal to 1.5 times the nominal diameter of the pipe. In this work, since 10 inch pipeline ID is adopted, the bending radius of skid 2 is 0.385 m.

3 Results and Discussion

3.1 Maneuverability of the Two-Module System

As mentioned previously, wheels are attached to ease the mobility of the prototype in a straight pipe as well as in a S-bend pipe, respectively. Similarly, the PU rubber is removed to avoid severe friction and damage since no fluid domain involved. However, the system is stuck at the S-bend and no movement can be made afterwards. The upper cut-out of the pipe allows the removal of the stuck capsule easily as described in Fig. 6. Thus, the PU cup and guiding discs are installed back to test the maneuverability and it works. The installation of PU with particular compression properties of rubber eases the mobility at bend and operation is successfully resumed as shown in Fig. 7. The PU cup compresses as it enters the pipeline due to its larger diameter than the inner diameter of the pipe. Therefore, the compression is likely to be critical at the bend area. In order to ensure the cup compresses adequately without exhibiting its yielding threshold, the following condition must be met in order to avoid failure of the cup.

$$\text{PUCup Equivalent Stress} < \text{PU Tensile Yield Stress(45.2MPa)}$$

Fig. 6 Insertion of prototype inside S-bend pipe and its consequence

Fig. 7 Installation of PU rubber and its execution

The value of the PU tensile yield stress is obtained from the ANSYS's library [5]. It is a mechanical property which measures the maximum amount of tensile stress a material can withstand before it undergoes nonlinear deformation under tension. In the context of polyurethane, it refers to the maximum amount of the tensile stress the material can hold before it fails. Thus, the PU integrity assessment by measuring tensile yield stress should be done prior industrial execution to evaluate its suitability for numerous application. Figure 8 illustrates that at bend radius 1.5D for the 25.4 cm pipe size, the maximum compression required is 25.67 mm which is measured from the engineering drawing. Moreover, it also illustrates that at maximum compression, the maximum equivalent stress experienced by the cup is 2.48 MPa which is much lower than the tensile yield stress which is 45.2MPa. Therefore, the PU cup design is safe by 18 factor of safety. In conclusion, the execution of experiment is carried out by attaching the wheels onto the prototype using skid 1 and PU rubber at skid 2, respectively.

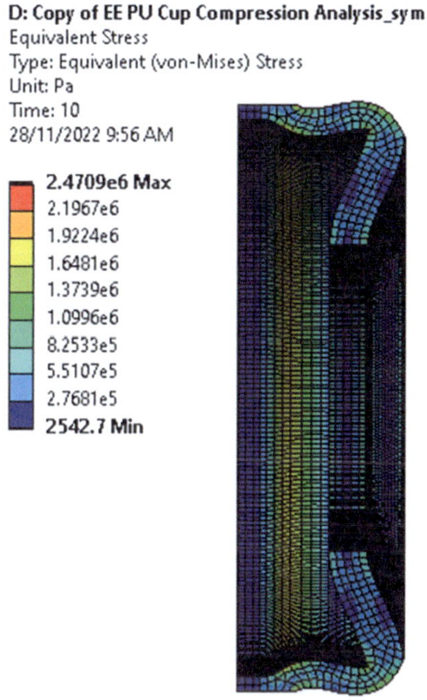

Fig. 8 Mechanical mobility of prototype inside the critical bend

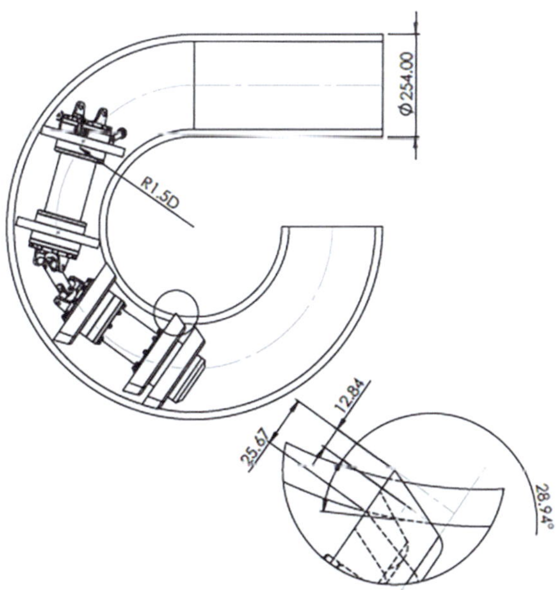

3.2 Detection of Radioactive Signals

The objective of this study is to verify the ability of the said system to detect the radioactive signal which is emitted from a small concentration of radioactive source. In this study, Cs137 is emitting gamma ray and SiPM is able to pick up the ionizing radiation not only in its stationary condition but also when it is mobilized. Detection of radioactive signals are conducted successfully in which the mobilized system is able to acquire the radioactive data from very small strength of activity as shown in Fig. 9. On top of that, the slit or the aperture of collimator which shields the radioactive source is in size of 5 mm. Thus, we can conclude that the novel pigging system is able to detect not only smaller concentration of leak but smaller in size as well. It also describes the speed traveled by the module inside the pipeline by adopting peak to peak method as shown in Eq. (1) [1, 6]:

$$v\left(\frac{m}{s}\right) = \frac{\Delta d\,(m)}{\Delta t(s)} \quad (1)$$

where v is the velocity (m/s), Δd is the distance between peaks in meter (m) and Δt is the time interval between two radioactive sources. Therefore, the measured speed of the system in this study is 0.14 m/s which is below the industrial range speed which is 1–3 m/s [7]. However, the works will be further improved in which some modification will be conducted to the winch system so that the range of industrial speed can be achieved. Future works should be conducted for various speeds of the said system and the distance's issue should be addressed prior embarking its utilization in actual industrial application.

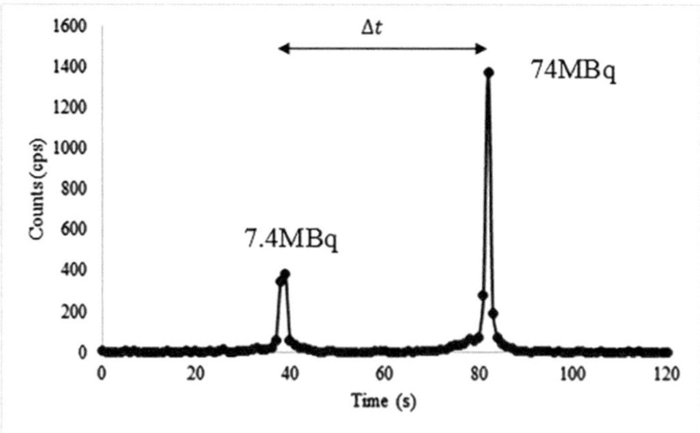

Fig. 9 Detection of Cs-137 with respect to distance

4 Conclusion

Several key takeaways can be extracted from this work:

1. A novel leakage detection system is complimentary with the conventional radiotracer technique has been developed and verified
2. The pigging system using SiPM as detector is able to acquire radioactive signals emits by Cs137 as radioactive material
3. The developed pigging system is capable to detect small and low concentration of radioactive signals which can be implied as capable to pick up low concentration and small size of leak for industrial utilization.

Acknowledgements This work is supported by Dana Pemacu Teknologi Strategik RMK 12, Ministry of Science, Technology and Innovation of Malaysia (MOSTI).The authors would like to extend their greatest appreciation to all colleagues from Prototype and Design Centre (PDC), Engineering Department (Electrical and Instrumentation), Engineering Department (Civil and Mechanical) and Plant Assessment Technology (PAT) from Malaysian Nuclear Agency who have directly and indirectly involved in this work and ensuring the fruitful execution of the project.

References

1. International Atomic Energy Agency, Radiotracer technology as applied to industry. IAEA TECDOC-1262 (2001). https://www-pub.iaea.org/MTCD/Publications/PDF/te_1262_prn.pdf. Accessed 6 Jun 2022
2. International Atomic Energy Agency, Leak detection in heat exchangers and underground pipelines using radiotracers. IAEA-TCS-38 (2009). https://www-pub.iaea.org/MTCD/Publications/PDF/TCS-38_web.pdf. Accessed 2 Feb 2024
3. N. Othman, N. Yussup, A. Z. Johan, J. Z. A. Rahman, R. Mostapa, Investigation of leakage in underground pipeline using radiotracer technology, in *Advancements in Materials Science and Technology Led by Women. Advanced Structured Materials*, vol. 16, ed. by A. Ismail, F. Nur Zulkipli, H. S. Husin, A. Öchsner (Springer, Cham). https://doi.org/10.1007/978-3-031-219599-7_3
4. S. El-Zahab, T. Zayed, Leak detection in water distribution networks: an introductory overview. Smart Water **4**(5) (2019) https://doi.org/10.1186/s40713-019-0017-x
5. M. Ashbey, *Material property data for engineering materials*, 5th edn (ANSYS Education Resources) (2021). https://www.ansys.com/content/dam/amp/2021/august/webpage-requests/education-resources-dam-upload-batch-2/material-property-data-for-eng-materials-BOKENGEN21.pdf. Accessed 13 April 2024
6. International Atomic Energy Agency, Residence time distribution method for industrial and environmental applications. IAEA-TCS-31. https://wwwpub.iaea.org/MTCD/Publications/PDF/TCS-31_web.pdf.(2008) Accessed 16 Feb 2024
7. F. L. Gustavo, C. G. Victor, R. P. Freitas, Maitelli A. L., A. O. Salazar, PIG's speed estimated with pressure transducers and hall-effect sensor: An industrial application of sensors to validate a testing laboratory. Sensors **17**(9), 2119 (2017). https://doi.org/10.3390/s17092119

A Comparative Total Cost of Ownership Analysis of Diesel Tugboat, Dual Fuel Tug Boat, and Electric Tug Boat for LNG Carrier Escorting Activities in Indonesia

Erzad Iskandar Putra, Wasis Dwi Aryawan, Ahmad Nasirudin, Danu Utama, and Md. Redzuan Zoolfakar

Abstract This study presents a comprehensive economic comparison of three distinct tugboat types: diesel, dual fuel LNG, and electric tugboats, with a particular focus on cost-competitiveness. The evaluation revolves around the total cost of ownership (TCO) value, encompassing capital expenditure (CAPEX), operating expenditure (OPEX), and voyage costs associated with each tugboat. Through a parametric study, the impact of tugboat utilization per year on the TCO value is meticulously analyzed. Simulating three utilization scenarios: 60 days, 120 days, and 149 days per year, the results reveal that the diesel tugboat exhibits the lowest TCO value for 60 days of operation annually. On the other hand, the dual fuel (LNG) tugboat showcases superior cost-effectiveness for 120 and 149 days of operation per year. These findings greatly contribute to informed decision-making processes, providing valuable insights into the cost-competitiveness of different tugboat options.

Keywords Cost-competitive · Total cost of ownership · Electric tugboat

---

E. I. Putra (✉) · W. D. Aryawan · A. Nasirudin · D. Utama
Department of Naval Architecture, Faculty of Marine Technology, Sepuluh Nopember Institute of Technology, Surabaya, Indonesia
e-mail: erzad@its.ac.id

W. D. Aryawan
e-mail: wasis@its.ac.id

A. Nasirudin
e-mail: anasirudin@its.ac.id

D. Utama
e-mail: danuutama@na.its.ac.id

Md. Redzuan Zoolfakar
Malaysian Institute of Marine Engineering Technology, Universiti Kuala Lumpur, Jalan Pantai Remis, Lumut, Perak, Malaysia
e-mail: redzuan@unikl.edu.my

A. Ismail et al. (eds.), *Advancements in Industry and Technology*,
SpringerBriefs in Applied Sciences and Technology,
https://doi.org/10.1007/978-3-031-97931-6_9

1 Introduction

The maritime shipping industry plays a vital role in global trade, and it comprises approximately 80–90% of the total international trade [1]. However, its significant reliance on fossil fuels has made it a significant contributor to global warming and climate change. The burning of heavy fuel oils by ships releases substantial amounts of greenhouse gases (GHG) into the atmosphere, including SOx, NOx, CO, CO_2, and PMs is produced as a byproduct [2]. The GHG emissions originating from maritime transportation account for approximately 3% of the total global emissions [3].

The International Maritime Organization (IMO) published a regulation with the aim of reducing harmful sulfur oxides (SOx) emissions by the maritime industry. The regulation was implemented on January 1, 2020, and requires ships to use fuel oil with a sulfur content of no more than 0.5% m/m (mass by mass) [4]. The enforcement of IMO 2020 is ensured by the 2019 Guidelines for Consistent Implementation of the 0.5 percent Sulphur Limit under MARPOL Annex VI [5].

There are two main categories of emission reduction technologies: primary methods that are engine-related and secondary methods that are engine retrofits (such as scrubbers, SCR, and waste heat recovery) technologies [6]. The installation of these emission reduction systems results in the utilization of additional space within the engine room and funnel of the ships. In the case of smaller work vessels like tugboats, where these areas are constrained and hold significant value, there are difficulties associated with incorporating these systems onto such ships.

The shipping industry is exploring a range of clean fuel alternatives to reduce emissions. Liquefied natural gas (LNG) as a fuel for marine vessels is a significant alternative that is projected to be the preferred choice among alternative fuels for new ships [7]. LNG as fuel results in the elimination of SOx emissions and significant reduction of NOx, CO_2, and particulate matter (PM) emissions when compared to the emissions from a typical vessel powered by marine diesel oil [8]. LNG can offer significant cost savings compared to traditional marine fuels. Another increasingly popular alternative fuel is the utilization of a battery system as the energy source for ship propulsion. The development of this battery system poses challenges due to the high costs involved in constructing ships equipped with this technology. Currently, only a limited number of ships, particularly tugboats, employ this system. Nevertheless, in 2022, Damen successfully launched its first electric tugboat, which is currently operational in the port of Auckland [9].

The techno-economic study of diesel and dual fuel (diesel and LNG fuel) has been investigated, and the results show that the dual fuel tugboat has a significantly higher capital expenditure compared to the diesel tugboat [10]. However, the operational expenditure of dual fuel tugboat is seen to be 5% to 55% less costly due to the price of LNG being much cheaper than diesel oil. In this study, the capital expenditure and operational expenditure of electric tugboats were made and compared with diesel and dual fuel (diesel-LNG) tugboats. Then, a parametric study was conducted to assess the impact of tugboat utilization, on an annual basis throughout its economic lifetime, on both the investment and operational costs of the tugboat.

2 Methodology

2.1 *Structure of the Total Cost of Ownership Model*

Total cost of ownership (TCO) analysis is a common practice for businesses to compare the costs and potential benefits associated with any action or decision. It serves as a useful tool for shipowners or investors to compare the costs of different types of vessels and identify the most economical option. TCO compares one single figure on a like-for-like basis to simplify the complex procedure of comparing multiple costs [11]. This process involves summing capital, operating expenditures, voyage cost and dividing by the economic lifetime of the vessel (Fig. 1).

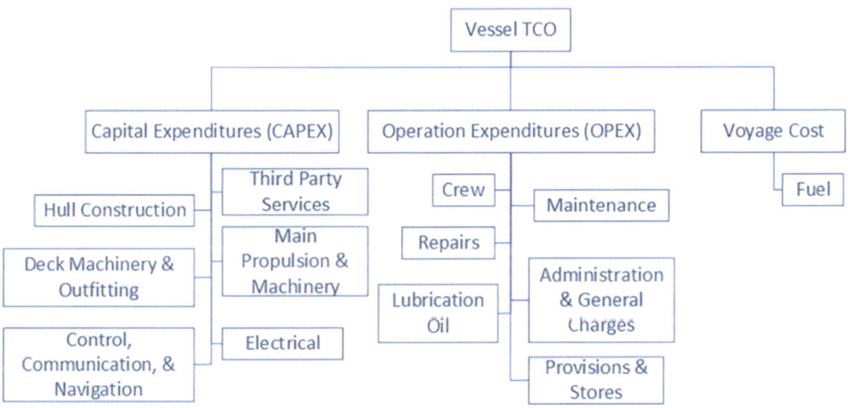

Fig. 1 Breakdown of the TCO components

2.2 *Capital Expenditures*

The capital expenditure (CAPEX) is the constant annual instalment to which the initial investment cost (IC) is equally distributed throughout the investment lifetime (n) under a determined discount rate (R) and can be calculated by the following equation [12];

$$\text{CAPEX} = \text{IC} \cdot \frac{R(1+R)^n}{(1+R)^n - 1} \tag{1}$$

2.3 Operation Expenditure

The operation expenditure (OPEX) is the sum of annual crew cost (ACC), annual maintenance cost (AMC), annual repair cost (ARC), annual administration and general charges (AAGC), annual lubrication oil cost (ALOC), annual provision and stores cost (APSC), and insurance cost (IC).

$$\text{OPEX} = \text{ACC} + \text{AMC} + \text{ARC} + \text{AAGC} + \text{ALOC} + \text{APSC} + \text{IC} \tag{2}$$

2.4 Voyage Cost

The voyage cost is the annual fuel cost (AFC) generated by the fuel oil price of diesel oil, LNG, and electricity power consumption for diesel tugboat, dual fuel tugboat, and electric tugboat, respectively.

$$\text{AFC} = \text{FP} \cdot \sum_{i=1} (\text{SFOC}_{\text{ME}} \cdot P_{\text{ME}} \cdot \text{RH}_{\text{ME}} + \text{SFOC}_{\text{AE}} \cdot P_{\text{AE}} \cdot \text{RH}_{\text{AE}}) \tag{3}$$

where;

SFOC_{ME}: Main engine specific fuel oil consumption [gr/kWh].
P_{ME}: Main engine power [kWh].
RH_{ME}: Main engine running hours [h].
SFOC_{AE}: Auxiliary engine specific fuel oil consumption [gr/kWh].
P_{AE}: Auxiliary engine power [kWh].
RH_{AE}: Auxiliary engine running hours [h].

3 Results and Discussion

3.1 Tugboat Specifications

In this case study, the dimensions, bollard pull, and propulsion system configurations of investigated tugboats are selected according to the existing tugboats. The short specification of investigated tugboats can be seen in Table 1.

Table 1 Selected tugboats short specifications

Properties		Tugboats of case study		
		Diesel	Dual fuel (LNG)	Electric
Main dimensions	Length	24.4 m	35 m	24.73 m
	Breadth	11.25 m	15.4 m	13.13 m
	Draft	5.6 m	5.5 m	4.95 m
	Speed	13 knots	13.5 knots	12 knots
	Bollard Pull	70 tons	70 tons	70 tons
Machinery	Main Engine	2 × 2100 kW 1600 rpm Caterpillar 3516C	2 × 1665 kW 1200 rpm Wartsila 9L20DF	2 × Ramme TW 1400r_L
	Auxiliary Engine	2 × 86 ekW Caterpillar C4.4	2 × 240 kW Nordhavn/Scania Diesel	2 × Caterpillar C32TA, 1175 kVA, 690 V, 60 Hz

3.2 CAPEX Estimation

In general, there are three stages of the estimation process, namely preliminary, intermediate, and final estimates. This study presents the estimation results of the tugboat CAPEX at the preliminary stage. The results of the estimated capital expenditure (CAPEX) can be seen in Table 2, which includes a shipyard profit margin of 10% and a tax rate of 11%. To calculate the annual instalment value of the CAPEX, an assumption of a 30% equity and 70% loan ratio is utilized. The loan interest rate applied is 8.5% per year, with a loan term of 10 years, considering the ship's economic lifespan of 25 years.

Table 2 Tugboats CAPEX estimation (in USD)

CAPEX item	Diesel	Dual fuel (LNG)	Electric
Third-party services	319,100	320,600	611,000
Hull construction	501,840	511,900	520,100
Main propulsion and machinery	1,556,440	2,586,400	6,329,060
Deck machinery and outfitting	1,187,320	1,210,200	1,714,180
Electrical	118,200	202,300	1,629,600
Control, communication, and navigation	139,100	141,400	344,200
Total construction cost	3,822,000	4,972,800	11,148,140
Profit (10%)	382,200	497,280	1,114,814
Tax (11%)	420,420	547,008	1,226,295
Grand total	4,624,620	6,017,088	13,489,249

3.3 Operation Expenditure

a. Annual Crew Cost
ACC includes all direct and indirect charges incurred by the crewing of the vessel. In this estimate, the annual crew costs for diesel tugboats is USD 201,342 and dual fuel (LNG) tugboats are assumed to be the same. However, for electric tugboats, the crew costs are slightly higher (USD 218,868) due to the requirement for crew members to possess advanced skills and qualifications to operate these types of tugboats.

b. Annual Maintenance Cost
Maintenance costs are assumed to increase based on the assumption of an inflation rate of 3%. The inflation rate is taken based Asian Development Bank (ADB) data forecast for Indonesia economy. The annual maintenance cost (AMC) for different types of tugboats are: USD 67,379 for diesel, USD 107,807 for dual fuel LNG, and USD 26,952 for electric.

c. Annual Repair Cost
Repair costs are assumed to increase based on the assumption of an inflation rate of 3%. The estimated annual repair cost (ARC) for different types of tugboats are: USD 111,836 for diesel, USD 178,937 for dual fuel LNG, and USD 106,244 for electric.

d. Annual Administration and General Charges
The estimated annual administration and general cost (AAGC) for different types of tugboats are: USD 27,748 for diesel, USD 28,025 for dual fuel LNG, and USD 28,025 for electric.

e. Annual Lubrication Oil Cost
The lubrication oil consumption is assumed at 0.35 g/kWh and the lubrication oil price at USD 2,646/ton. On an electric tugboat, lubricant consumption is only used for auxiliary engines. The estimated annual lubrication oil cost for a diesel tugboat is USD 1,560, for a dual fuel LNG tugboat is USD 1,560, and for an electric tugboat is USD 1,092.

f. Annual Provision and Stores Cost
The APSC is assumed to be the same for the three types of ships at USD 20,811 per year. Provision and stores cost are assumed to increase based on the assumption of an inflation rate of 3%.

g. Insurance Cost
In this study, the insurance cost per year is 0.7% of the ship's price (CAPEX). The estimated annual insurance cost for a diesel tugboat is USD 32,372, for a dual fuel LNG tugboat is USD 42,120, and for an electric tugboat is USD 74,933.

3.4 Voyage Cost

Voyage costs refer to the fuel expenses incurred during the operations of tugboats. The duration of tugboat operations is assumed to be eight hours per day, with a total

Table 3 Tugboat fuel price

Fuel type	Price	Unit
Diesel oil (HSD B30)	IDR 15,000 / USD 0.96	Per liter
LNG	IDR 186,000.00 / USD 12	Per MMBtu
Electricity	IDR 1680 / USD 0.11	Per kWh

Table 4 Annual tugboat voyage cost for 60 days/year (USD)

Voyage cost	Diesel	Dual fuel (LNG)	Electric
Fuel cost	361,417	179,459	54,627

of 60 operating days per year. These operation durations are based on estimated port calls per year at a private port in Indonesia. The fuel price for tugboat operation can be seen in Table 3. The fuel consumption for a diesel tugboat is 199.5 g/kWh or 5.6 tons/day (8 h of operation). While the dual-fuel ASD tug LNG gas consumption is 0.007876359 MMBtu/kWh or 221.17 MMBtu/day (8 h operation). Diesel oil in the dual-fuel ASD tug is only used at the initial start of the engine and the rest uses LNG gas. The comparison of the use of diesel fuel with LNG gas for a dual-fuel ASD Tug is 1:9. The estimation result of fuel cost for 60 days operation in a year can be seen in Table 4.

Battery replacement costs are exclusive to electric tugboats, occurring every 10 years of operation at an estimated cost of approximately USD 700,000 per replacement. Hence, the annual battery replacement cost is estimated to be around USD 56,000 for the entire operating year.

3.5 *Tugboat Utilization Parametric Study*

A parametric study was conducted to examine the impact of tugboat utilization per year on the total cost of ownership (TCO) value. The study simulated three scenarios of tugboat utilization: 60 days, 120 days, and 149 days per year. Table 5 presents the capital expenditure (CAPEX), operating expenditure (OPEX), and their relationship to utilization days (60, 120, and 149 days) for the three types of tugboats.

Table 5 Tugboat TCO result

Variables	Diesel engine	Dual fuel (LNG)	Electric
Operating cost			
Fuel consumption			
– 60 operating days/year	361,417	179,459	54,627
Annual cost saving	0%	50%	85%
– 120 operating days/year	722,834	358,917	109,254
Annual cost saving	0%	50%	85%
– 149 operating days/year	897,518	445,656	135,657
Annual cost saving	0%	50%	85%
OPEX			
– 60 operating days/year	824,465	760,061	587,552
Annual cost saving	0%	8%	29%
– 120 operating days/year	1,187,442	941,080	643,271
Annual cost saving	0%	21%	46%
– 149 operating days/year	1,362,881	1,028,572	689,694
Annual cost saving	0%	25%	49%
Capital cost			
CAPEX			
- **Purchase price**	**USD 4,625,000**	**USD 6,017,000**	**USD 13,489,249**
Total cost of ownership			
TCO			
– 60 operating days/year	**USD 10,771,660**	USD 11,255,375	USD 16,393,582
– 120 operating days/year	USD 12,958,866	**USD 12,576,396**	USD 16,798,949
– 149 operating days/year	USD 14,705,731	USD **13,768,888**	USD 16,994,877

4 Conclusion

The results show that the electric tugboat saves 29% of annual OPEX compared to the diesel engine ASD tug when operating for 60 days per year. With 120 days of utilization per year, the cost savings increase to 46%. Furthermore, with 149 days of utilization per year, the cost savings slightly increase to 49%.

Based on the TCO results, the diesel tugboat has the lowest TCO value compared to the dual fuel (LNG) tugboat and electric tugboat when operating for 60 days per year. However, if the utilization days increase to 120 days per year or 149 days per year, the dual fuel (LNG) tugboat exhibits the lowest TCO value. Notably, for 149 operating days, both the diesel tugboat and electric tugboat show equal TCO values.

Acknowledgements This research was funded by the Department of Naval Architecture, Faculty of Marine Technology, Institut Teknologi Sepuluh Nopember research scheme.

References

1. P. Balcombe, J. Brierley, C. Lewis, L. Skatvedt, J. Speirs, A. Hawkes, I. Staffell, How to decarbonise international shipping: options for fuels, technologies and policies. Energy Convers Manag. **182**, 72–88 (2019). https://doi.org/10.1016/j.enconman.2018.12.080
2. C. Deniz, B. Zincir, Environmental and economical assessment of alternative marine fuels. J. Clean. Prod. **113**, 438–449 (2016). https://doi.org/10.1016/j.jclepro.2015.11.089
3. S. Singh, V. Rambarath-Parasram, Considerations for the Latin American and Caribbean region in light of the global move towards low carbon shipping. in *Best Practices in Manufacturing Processes* (Springer International Publishing, Cham, pp. 53–73) (2019)
4. International Maritime Organization: IMO 2020 – cutting sulphur oxide emissions
5. Marine Environment Protection COMMITTEE (MEPC): 2019 Guidelines for consistent implementation of the 0.50% sulphur limit under Marpol Annex VI., London (2019)
6. S. Lion, I. Vlaskos, R. Taccani, A review of emissions reduction technologies for low and medium speed marine Diesel engines and their potential for waste heat recovery. Energy Convers Manag. **207**, 112553 (2020). https://doi.org/10.1016/j.enconman.2020.112553
7. F. Burel, R. Taccani, N. Zuliani, Improving sustainability of maritime transport through utilization of Liquefied Natural Gas (LNG) for propulsion. Energy **57**, 412–420 (2013). https://doi.org/10.1016/j.energy.2013.05.002
8. M. Pawlak, Analysis of Economic Costs and Environmental Benefits of LNG as the Marine Vessel Fuel. Solid State Phenom. **236**, 239–246 (2015). https://doi.org/10.4028/www.scientific.net/SSP.236.239
9. Rick van de Weg: Damen's first all-electric tug Sparky delivered to Ports of Auckland
10. Ö.E. Karaçay, O.A. Özsoysal, Techno-economic investigation of alternative propulsion systems for tugboats. Energy Convers. Manage. **X** 12, (2021). https://doi.org/10.1016/j.ecmx.2021.100140
11. C. Rout, H. Li, V. Dupont, Z. Wadud, A comparative total cost of ownership analysis of heavy duty on-road and off-road vehicles powered by hydrogen, electricity, and diesel. Heliyon. **8**, e12417 (2022). https://doi.org/10.1016/j.heliyon.2022.e12417
12. G.A. Livanos, G. Theotokatos, D.-N. Pagonis, Techno-economic investigation of alternative propulsion plants for Ferries and RoRo ships. Energy Convers Manag. **79**, 640–651 (2014). https://doi.org/10.1016/j.enconman.2013.12.050

Wireless Communication Street Lights via Long Range Technology

Mohd Falfazli Mat Jusof, Mohd Helmi Suid, Mohd Ashraf Ahmad, Taharah Abdullah, and Muhamad Husaini Abu Bakar

Abstract Street lighting is a vital component for urban safety and functionality, requiring regular maintenance to ensure optimal performance. This research explores the use of the long range (LoRa) technology in a mesh network topology for efficient streetlight monitoring and fault detection. By leveraging LoRa's low-power, long-range communication capabilities, the system allows data transmission between streetlights over distances of up to 240 m. The mesh topology enhances the network's reliability by enabling nodes to receive and retransmit data, ensuring robust communication even at extended distances. This approach facilitates real-time monitoring of streetlight conditions, enabling the remote detection of faults and the prioritization of maintenance tasks. The proposed system significantly reduces the time and resources needed for maintenance, providing an efficient solution for managing large urban streetlight infrastructures. The findings confirm the effectiveness of LoRa technology and mesh networking in enhancing streetlight monitoring systems, streamlining maintenance workflows, and improving overall service efficiency.

Keywords Streetlight monitoring · LoRa technology · Mesh network topology · Automated maintenance system · Real-time monitoring · Fault detection

M. F. M. Jusof (✉) · M. H. Suid · M. A. Ahmad · T. Abdullah
Faculty of Electrical and Electronics Engineering Technology, Universiti Malaysia Pahang Al-Sultan Abdullah, Kuantan, Malaysia
e-mail: mfalfazli@umpsa.edu.my

M. H. Suid
e-mail: mhelmi@umpsa.edu.my

M. A. Ahmad
e-mail: mashraf@umpsa.edu.my

T. Abdullah
e-mail: taharahabdullah96@gmail.com

M. H. A. Bakar (✉)
System Engineering and Energy Laboratory, Malaysian Spanish Institute, Universiti Kuala Lumpur, Kulim Hi-Tech Park, Kulim, Kedah, Malaysia
e-mail: muhamadhusaini@unikl.edu.my

A. Ismail et al. (eds.), *Advancements in Industry and Technology*,
SpringerBriefs in Applied Sciences and Technology,
https://doi.org/10.1007/978-3-031-97931-6_10

1 Introduction

Street lighting is vital for public safety, security, and accessibility in urban and rural areas [1, 2]. In Malaysia, streetlights reduce accidents, deter crime, and support daily activities [3]. However, maintaining their functionality requires regular inspections and timely repairs. The current reactive maintenance approach relies on public complaints and manual inspections, leading to delays, increased costs, and service outages.

To address these challenges, this work proposes an automated streetlight monitoring system using the long range (LoRa) technology and a mesh network topology. LoRa, a low-power wide-area network (LPWAN) protocol, enables long-range, low-power communication, ideal for monitoring devices over large areas [4, 5]. The mesh network ensures robust data transmission, even in areas with limited connectivity. This system allows real-time monitoring, immediate fault detection, and prompt maintenance, improving efficiency and reducing costs.

Key Contributions:

1. Automated Fault Detection: Replaces reactive maintenance with proactive, automated solutions.
2. Advanced Communication: Integrates LoRa and mesh networking for energy-efficient, long-range communication.
3. Scalability: Designed for large-scale streetlight networks in urban and rural areas.
4. Enhanced Safety and Resource Management: Improves reliability and optimizes maintenance workflows.

This system enhances streetlight maintenance, ensuring reliable, efficient, and cost-effective public lighting infrastructure.

2 Methodology

This section outlines the methodology used for the design, development, and implementation of the wireless communication system for streetlights using the LoRa technology. The approach integrates both hardware and software components to create a reliable and efficient system for monitoring and controlling streetlights. The primary goal is to develop a wireless communication system that can monitor and control streetlights using the ESP32 microcontroller integrated with the LoRa technology. The system is designed to be low-power, scalable, and capable of operating over long distances with minimal data transmission. The wireless network enables communication between streetlight nodes and a central control office, where the status of each light can be monitored and controlled.

2.1 Hardware Requirements

Microcontroller Selection and LoRa Technology

The ESP32 microcontroller, integrated with LoRa technology, is chosen for its long-range communication and low-power consumption [6]. It features a 32-bit processor, GPIO pins for sensor connections, and built-in Wi-Fi and Bluetooth. For this project, the ESP32 is enhanced with LoRa modules (TTGO LoRa SX1278 and Heltec LoRa SX1276), which include built-in LoRa transceivers. External antennas are required for proper transmission, and care must be taken to connect them before uploading code to avoid damage.

LoRa, a wireless communication technology, uses radio frequency modulation to enable long-range, low-power communication [5]. Operating at 868 and 915 MHz, it is ideal for IoT applications requiring wide-area coverage and minimal data transmission [7, 8]. This makes it suitable for remote streetlight monitoring, as illustrated in Fig. 1.

Software Development

The ESP32 is programmed using the Arduino IDE (C language) to configure the LoRa transceiver, set up communication protocols, and implement frequency settings (868 MHz or 915 MHz). Each ESP32 is programmed for two-way communication, enabling data exchange between streetlight nodes. A mesh network topology ensures robust data transmission, where nodes communicate directly or via intermediate nodes, enhancing reliability.

The user interface (UI) is developed using Visual Studio IDE with Visual Basic.NET. It allows the control center to:

- Monitor real-time streetlight status (on/off, faults).
- Remotely control individual or groups of streetlights.
- Display fault conditions and system alerts.

System Workflow

1. Data Collection: Each streetlight node (equipped with ESP32 and LoRa) collects operational data (e.g., on/off status, faults).

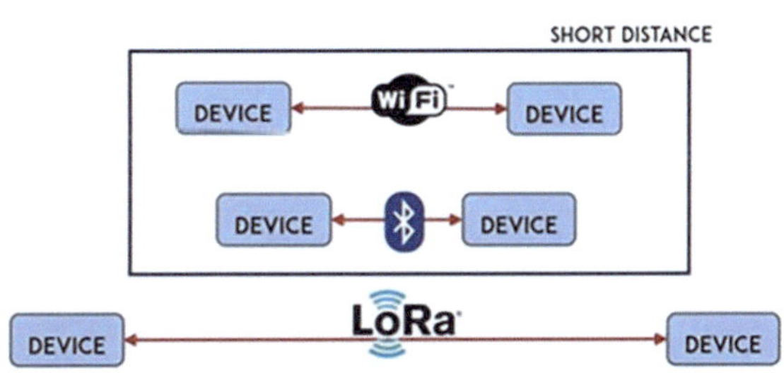

Fig. 1 Illustration of LoRa technology for long-range communication

2. Data Transmission: Data is transmitted via LoRa to nearby nodes, forming a network that relays information to the central control office.
3. Control Office Interface: The office receives and displays real-time data on the UI for monitoring and management.
4. Control Commands: The office can send commands (e.g., turn lights on/off, report faults) to individual nodes.

The complete design of the system, which facilitates this workflow, is shown in Fig. 2.

Implementation Flow

1. Hardware Setup: ESP32 microcontrollers with LoRa modules are integrated, antennas connected, and power sources configured as shown in Fig. 3.
2. Software Development: ESP32 programming and UI development are completed. A flowchart visualizes communication logic as seen in Fig. 4a of

Fig. 2 Complete design of system

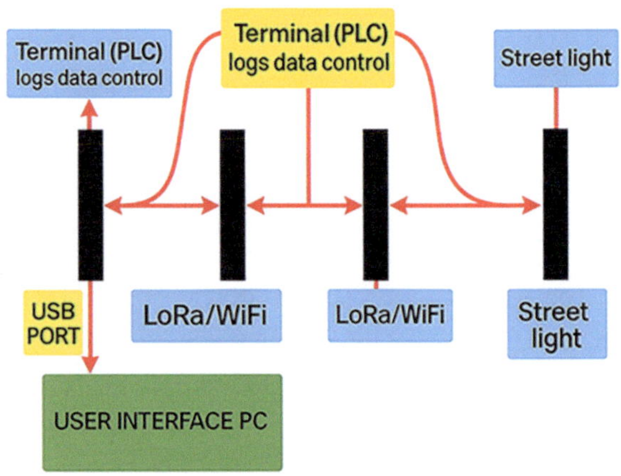

Fig. 3 ESP32 with LoRa technology

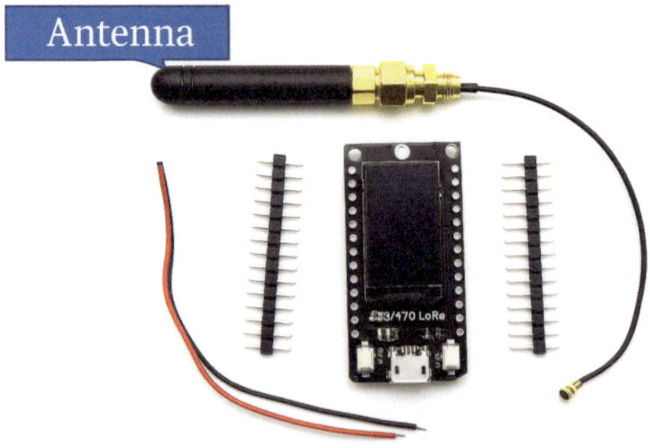

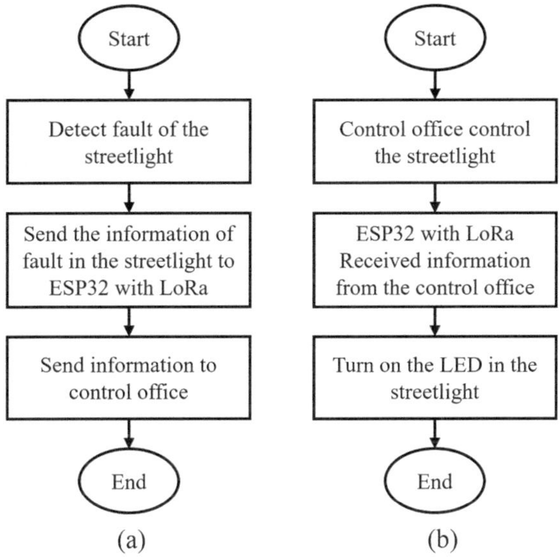

Fig. 4 Flowcharts for information exchange between streetlight and control office: **a** Flowchart sending the information from the streetlight, **b** Flowchart receiving the information from the streetlight

the flowchart shows how information is sent from the streetlight, and Fig. 4b shows how the control office receives and processes the data.

3. Testing: The system is tested to ensure proper communication, remote control, and fault detection. Debugging resolves any issues.
4. Validation: The system is validated for communication reliability, data transfer, and UI performance, with adjustments made as needed.

This system enhances streetlight monitoring, enabling real-time fault detection, remote control, and efficient maintenance, ensuring reliable and cost-effective public lighting infrastructure.

3 Results and Discussion

The following section provides a comprehensive analysis, and results obtained during the implementation of the research, beginning from the initial testing of basic communication to the final implementation of the monitoring system.

3.1 Basic Transmit and Receive Result

The initial phase tested basic data transmission between two ESP32 LoRa modules. Figure 5a shows the sender displaying packet ID and data, while Fig. 5b shows the receiver in standby mode, confirming transmission functionality. The next phase

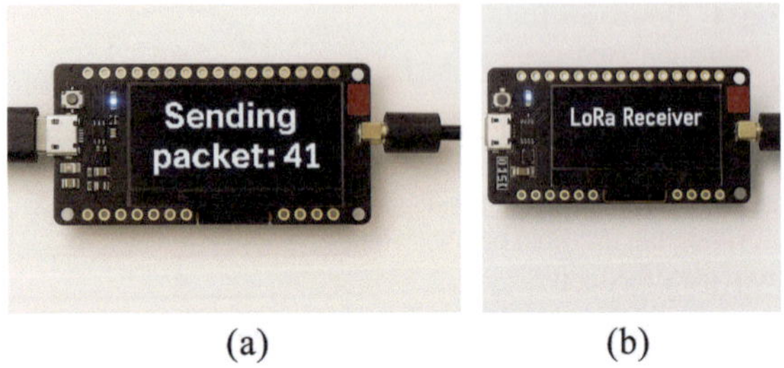

Fig. 5 **a** The first ESP32 for sending the information, **b** The second ESP32 before receiving the information

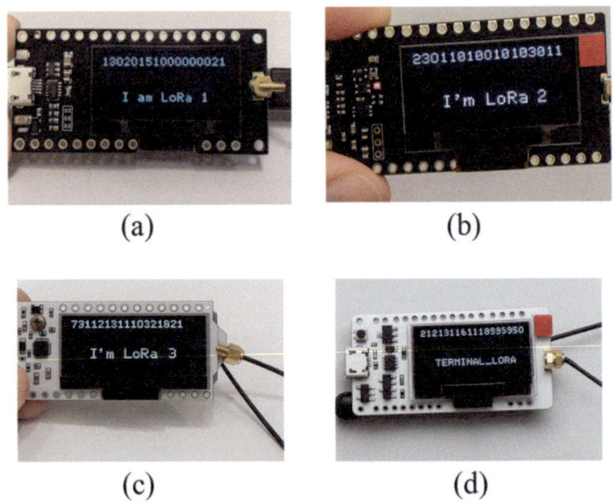

Fig. 6 **a** LoRa 1, **b** LoRa 2, **c** LoRa 3, **d** LoRa 4 (Terminal)

implemented a mesh network with three LoRa modules and a terminal (LoRa 4) connected to a user interface. Figure 6a–c demonstrate successful data reception among LoRa 1, 2, and 3, with values like "30," "20," and "31" indicating seamless communication. Figure 6d shows LoRa 4 aggregating data from all nodes, receiving values such as "21," "31," and "10," confirming robust communication and data aggregation across the mesh network. This setup validates the system's scalability and reliability for streetlight monitoring.

3.2 *Monitoring Result*

The streetlight system's monitoring and control were enabled through a user interface that displayed data from all LoRa modules. Figure 7 shows the interface, with LoRa 1 and LoRa 2 marked as "good" (values "11" and "21"), while LoRa 3 indicates an issue (value "30"). The interface also allows users to control LED indicators, as shown in Fig. 8a–c, where LEDs for LoRa 1, 2, and 3 are turned on. This demonstrates real-time monitoring and control capabilities.

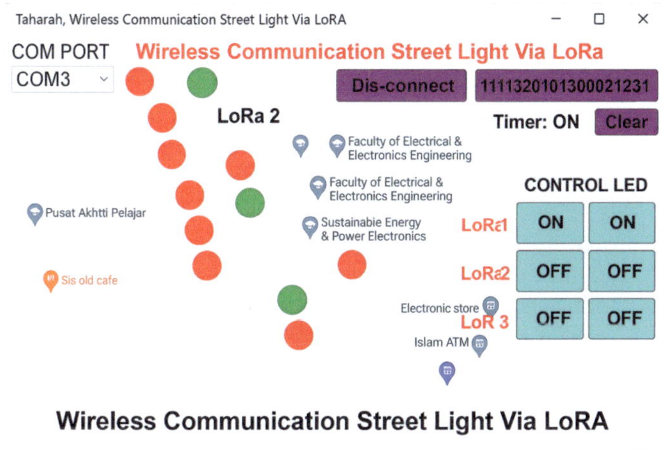

Fig. 7 User interface display

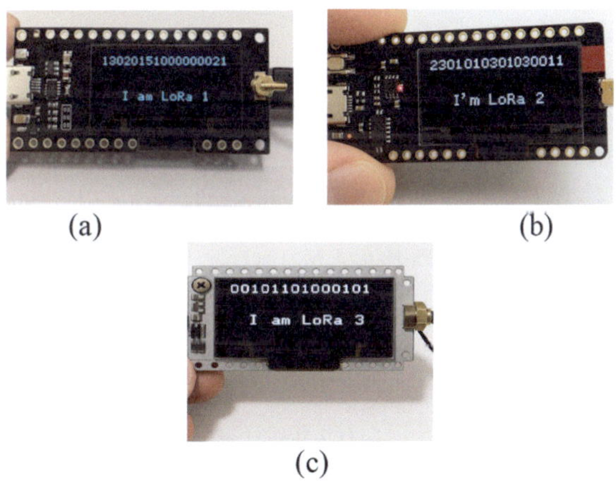

Fig. 8 **a** LoRa 1 controlled by the user interface (LED ON), **b** LoRa 2 controlled by the user interface, **c** LoRa 3 control by the user interface (LED ON)

In summary, the system successfully integrated LoRa technology, validated through one-way communication tests and a scalable mesh network. Data aggregation at the terminal module and the user interface's interactive features highlight the system's robustness and efficiency for smart city applications.

4 Conclusion

In this research, the effectiveness of LoRa technology for long-range, low-power wireless communication in streetlight monitoring systems was thoroughly investigated. By leveraging a mesh topology, the system extended the communication range, allowing reliable data transmission over distances of up to 240 m (approximately seven streetlight poles). The results demonstrated that LoRa could be effectively deployed in real-world applications, such as streetlight monitoring, where data needs to be transmitted over long distances in an urban environment. The system architecture, which incorporated ESP32 modules, allowed for the transmission of streetlight condition data between multiple modules in a chain. The communication distance was successfully measured and confirmed to be 240 m. The use of mesh networking ensured that data could be relayed over greater distances, with each module acting as a relay. Although there were slight delays in data transmission, the core objective of real-time monitoring and control of streetlight operations was still achieved, thereby confirming the system's practicality and reliability for smart city applications.

Acknowledgements We extend our deepest gratitude to University Malaysia Pahang Al-Sultan Abdullah for their substantial financial and resource support through the product development grant 2023 (PDU233213).

References

1. J. Green, C. Perkins, R. Steinbach, P. Edwards, Reduced street lighting at night and health: a rapid appraisal of public views in England and Wales. Health Place **34**, 171–180 (2015)
2. L. Bogdanov, S. Polstra, P. Yakimov, M. Marinov, Daedaled: a GUI tool for the optimization of smart city LED street lighting networks, in *2018 IEEE XXVII International Scientific Conference Electronics-ET*. IEEE, (2018), pp. 1–4
3. T.K. Nath, S.S.Z. Han, A.M. Lechner, Urban green space and well-being in Kuala Lumpur Malaysia. Urban For Urban Green **36**, 34–41 (2018)
4. M.S. Yahya, S. Soeung, S.K.A. Rahim, U. Musa, S.S.B. Hashwan, Z. Yunusa, S.A. Hamzah, LoRa microstrip patch antenna: a comprehensive review. Alex. Eng. J. **103**, 197–221 (2024)
5. L. Trinh, T.Q.K. Nguyen, D. Phan, V. Tran, V. Bui, N. Truong, F. Ferrero, Miniature antenna for IoT devices using LoRa technology, in *2017 International Conference on Advanced Technologies for Communications (ATC)*. IEEE (2017), pp. 170–173
6. A.A.H. Isa, M.T. Iqbal, Remote low-cost web-based battery monitoring system and control using LoRa communication technology. J. Electron. Electr. Eng. **2024**, 125–147 (2024)
7. Maier, A., Sharp, A., Vagapov, Y.: Comparative analysis and practical implementation of the ESP32 microcontroller module for the Internet of Things, in *2017 Internet Technologies and Applications (ITA)*. IEEE (2017), pp. 143–148
8. I. Allafi, T. Iqbal, Design and implementation of a low-cost web server using ESP32 for real-time photovoltaic system monitoring, in *2017 IEEE Electrical Power and Energy Conference (EPEC)*. IEEE, pp. 1–5

Classification of Customer Decision Factors Using Sea Transportation Mode Using the Gath-Geva Algorithm Method

Fitri Andy, Lucky Rasyad, and I Gusti Ngurah Sumanta Buana

Abstract Sea transportation is one of the choices for public travel facilities besides land and air transportation. Several factors, both internal and external, affect the decision-making process of individuals when selecting sea transportation as a travel option. Internal factors originate from the customer's personal preferences and circumstances, whereas external factors are influences that arise from outside the individual, playing a role in shaping their choices. The internal factors used in this study consisted of community expectations, the distance of residence to the port and the community knowledge of the services of the chosen ship company. The external factors consist of the number of crew members, the level of security, the availability of prayer facilities and also the discounted ticket prices given. In this study, these factors will be grouped to find out about people's decisions in choosing sea transportation as a means of travel using the Gath-Geva algorithm clustering method. The Gath-Geva algorithm builds upon the Gustafson-Kessel algorithm, offering the capability to identify clusters with a range of shapes, sizes, and densities. It utilizes fuzzy maximum likelihood estimates for distance measurements and incorporates an exponential component to enhance its clustering process. The results of this study are that the majority of users of sea transportation services use services at the port of Java island, namely with a percentage of 46.43% and the remaining 53.57% are outside Java island where people choose to use travel services via sea based on internal factors or factors that come from within themselves, namely service expectations with a classification accuracy of 62.44%.

Keywords Customer decisions · External factors · Fuzzy c-means · Gath-Geva algorithm · Internal factor

---

F. Andy (✉) · L. Rasyad · I. G. N. S. Buana
Institut Teknologi Sepuluh Nopember, Kampus ITS, Sukolilo, Jl. Raya ITS, Keputih, Surabaya, Kota SBY, Indonesia
e-mail: andy.gr4ss@gmail.com

L. Rasyad
e-mail: luckydrasyad@gmail.com

I. G. N. S. Buana
e-mail: buana@seatrans.its.ac.id

A. Ismail et al. (eds.), *Advancements in Industry and Technology*,
SpringerBriefs in Applied Sciences and Technology,
https://doi.org/10.1007/978-3-031-97931-6_11

1 Introduction

The growing need for daily activities heavily relies on land, sea, and air transportation. Particularly, sea transportation is a popular choice for areas separated by the ocean. Recognizing this, the Indonesian government has been actively constructing ports across all provinces. By 2020, Indonesia saw its ports increase to 2439, a 38.6% rise from 2019s 1760 ports. Additionally, to enhance shipping activities, pier construction reached 41,972 m in 2020, marking a 41.3% increase from the previous year. Sea transportation facilities also grew by 14.2% in 2020, totalling 93,737 units, including cargo ships, passenger ships, and fishing vessels [1]. This expansion in port facilities influences people's transportation choices, guided by both internal factors (like expectations, distance from port, and knowledge of shipping services) and external factors (such as crew numbers, security level, prayer facilities, and ticket discounts). This study employs the Gath-Geva algorithm for clustering to identify which factors, internal or external, most affect sea transportation choices. This algorithm helps in detecting various cluster shapes, sizes, and densities, providing insights for shipping companies to enhance services and meet customer needs effectively [2]. The cluster Gath-Geva algorithm clusters using membership functions or degrees and smoothes groups by not shifting the cluster center considerably [3] and groups outlier and overlapping instances [4].

2 Methodology

2.1 Clustering Assumption Testing

Classification using the clustering method can be carried out if the basic clustering assumptions are fulfilled:

1. Multivariate Normal Assumptions

   The assumption of multivariate normality is established to verify the normal distribution of data intended for analysis, based on the following hypothesis [5].

   H_0 = The data is normally distributed multivariate

   H_1 = The data is not normally distributed multivariate

   Visually, testing the multivariate normal assumptions can be seen using the *Q–Q* Plot. In a *Q–Q* plot, data results from a sample that is plotted in comparison with the expected quantiles from a normal distribution. If the sample data follows a normal distribution, the plot will show a linear diagonal line [6]. As for the mathematical way, it can be calculated by the formula,

$$\left(d_j^2, \chi_p^2\left(\frac{j-0.5}{n}\right)\right) \tag{1}$$

where d_j^2 is the value of the jth data distance and if it is generated p-value $> \alpha$, it can be concluded that the data is normally multivariate distributed [7].

2. Data Adequacy Assumptions

   The data adequacy test was carried out to find out whether the data used was sufficient to be classified using the clustering method used. In this study, the Kaiser-Meyer-Olkin (KMO) test with the hypothesis will be used [8].

   $H_0 =$ Data sufficient for analysis

   $H_1 =$ Data not sufficient enough for analysis

   If the results of KMO <0.5 are obtained, it can be concluded that H_0 is rejected which means that the data is not enough to be analyzed.

3. Multicollinearity Assumptions

   The purpose of the multicollinearity test is to identify if there is a significant correlation among independent variables [9]. Multicollinearity refers to the scenario where variables in a multiple regression model exhibit strong correlations with each other [10]. When a strong correlation exists among independent variables, it can interfere with the relationship between these variables and the dependent variable. Addressing this issue is crucial before proceeding. There are various methods to identify multicollinearity, which include:

   1. The value of the variance inflation factor (VIF) is more than 10.
   2. If there is a high correlation between the independent variables, which is greater than 0.90.
   3. When the coefficient of determination R^2 is high but there are no significant independent variables.

2.2 *Gath-Geva Algorithm*

The Gath-Geva algorithm evolves from the Gustafson-Kessel (GK) algorithm, capable of identifying clusters with diverse shapes, sizes, and densities. It employs fuzzy maximum likelihood estimates (FMLE) for calculating distances and incorporates an exponential component. The methodology of the Gath-Geva algorithm is outlined below.

1. Determine the initial initiation of the U partition matrix randomly and calculate the center of each cluster.
2. Calculate the distance between clusters

$$D_{ik} = \frac{(2\pi)^{\left(\frac{\pi}{2}\right)}\sqrt{\det(F_{\omega i})}}{\alpha_i} \exp\left(\frac{1}{2}(x_k - v_i)^T F_{\omega i}^{-1}(x_k - v_i)\right) \quad (2)$$

With,

$$F_{\omega i} = \frac{\sum_{k=1}^{N}\left(\mu_{ik}^{\omega}(x_k - v_i)(x_k - v_i)^T\right)}{\sum_{k=1}^{N}(\mu_{ik})^{\omega}} \text{ and } \alpha_i = \frac{1}{N}\sum_{k=1}^{N}\mu_{ik}$$

where,

x_k	is the k-th data point in the cluster.
v_i	the mean of data points in the i-th data cluster.
$F_{\omega i}$	fuzzy covariance matrix of the i-th cluster.
α_i	prior probabilities of the selected clusters.
μ_{ik}	the posterior probability of the i-th cluster selected.

3. Calculating the value of the new membership function (u_{t+1})

$$u_{i+k} = \left[\sum_{j=1}^{c} \left(\frac{D(x_k, v_i)}{D(x_k, v_j)} \right)^{\frac{2}{m-1}} \right]^{-1} \tag{3}$$

4. Return to step 2 if the change in the membership function value is still above
5. The threshold value $\|U_{t+1} - U_t\| < \varepsilon$

2.3 Validity Index

The validity index is a measure used to determine the optimal group. As for this research, the validity index used is,

1. Silhouette coefficient (SC)

   Silhouette is a method for evaluating or testing the quality of clusters by calculating the average dissimilarity between points in the same cluster and different clusters [11]. The highest silhouette value is the number of clusters selected.
2. Xie-Beni Index

This index aims to measure the ratio of the total variation in clusters and cluster separators with the following formula:

$$XB = \frac{\sum_{i=1}^{c} \sum_{k=1}^{N} \mu_{ik}^{m} \|x_k - v_i\|^2}{N\left(\min_{h \neq k} \|v_k - v_i\|^2\right)} \tag{4}$$

With,

N	The number of sample.
c	The number of cluster.
μ_{ik}	the k-th membership score indicates a better cluster partition.
m	fuzzifier.
$\|x_k - v_i\|$	Euclidean distance of data point to cluster center.
$\|v_k - v_i\|$	Euclidean distance between cluster centers.

A low Xie-Beni value indicates a better cluster partition.

2.4 Data Source

This research utilizes primary data gathered from a survey of 56 individuals who have traveled by sea transportation. The participants are categorized based on two travel routes: within Java island and outside Java island. The variables used to group factors that influence the community's use of sea transportation modes consisting of internal factors and external factors are the number of crew members (X_1), the percentage of service expectations (X_2), the distance between residence and the port (X_3), the percentage of security level (X_4), port location (X_5), and categories of factors that influence decisions (X_6). This research focuses on selecting sea transportation modes based on all shipping activities, such as passengers and goods. The types of goods focused on in this research based on the survey results are basic necessities such as rice, cooking oil, and clothing.

3 Results and Discussion

3.1 Data Characteristics

The characteristics of the research data can be identified using descriptive statistics which can provide general information about the research data to be analyzed. In this study, descriptive statistics were calculated for each continuous variable shown in Table 1. Table 1 provides information about the mean, variance, minimum and maximum values of each variable and it is known that the variance values of the variable percentage of service expectations (X_2) and percentage of security level (X_4) are higher than the other two variables indicating that these two variables have a data range that is quite heterogeneous compared to the other two variables, namely the number of crew members (X_1) and the distance between the residence and the port (X_3). Furthermore, it is known that the majority of port locations used by customers of sea transportation modes where these locations are categorized into two, namely outside Java and Java island presented in Fig. 1. The information obtained based on Fig. 1 is that the majority of users of sea transportation services use services at the port is not from the Java island, this can be because many industrial activities are centered on the island of Java so people use sea transportation to go to Java island.

Table 1 Descriptive statistics

Variable	Mean	Variance	Min	Max
X_1	12.696	8.543	8.000	27.000
X_2	75.390	142.900	50.000	91.000
X_3	1.625	0.530	1.000	3.000
X_4	75.23	126.110	45.000	90.000

Fig. 1 Passenger origin bar chart

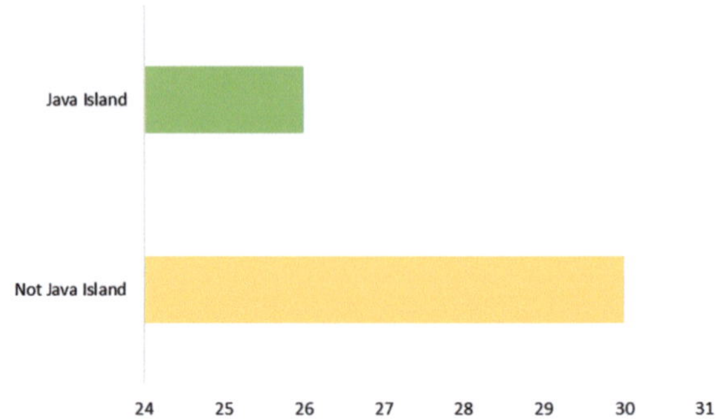

3.2 Test Clustering Assumptions

Clustering assumption testing is carried out so that in grouping there are no basic problems that can interfere with the results of the analysis. The several assumptions that must be met before clustering are,

1. Multivariate Normal Assumptions

The variables used in the study must have multivariate normal assumptions which can be seen using the Q-Q plot and also the correlation of the Mahalanobis distance (distance of the data from the cluster center) using the hypothesis,

H_0 = Data normally distributed multivariate

H_1 = Data not normally distributed multivariate

The *Q–Q* plots produced in this study are as follows presented in Fig. 2. Visually, the data used is a multivariate normal distribution because the plot between the Mahalanobis distance and the distance between clusters follows a linear line even though there are outlier data. However, to be sure, correlation testing can be carried out as shown in Table 2. The resulting correlation coefficient based on the output of Table 2 is 0.921 where this value indicates a high correlation and a p-value of <0.001 is obtained where this value is less than the significance level of 0.05 which means that it can be concluded that H0 is rejected. This means that there is a significant correlation and it indicates that the data has followed the normal multivariate distribution or the normal multivariate assumptions have been fulfilled.

2. Data Adequacy Assumptions

The Kaiser-Meyer-Olkin measure of sampling adequacy, known as the KMO MSA method, evaluates whether a data set is suitable for factor analysis by checking its adequacy for this type of analysis. In this research, a KMO MSA value of 0.511 was obtained, indicating that with a KMO value greater than 0.5, factor analysis is deemed appropriate, signifying that the data is sufficiently robust for analysis.

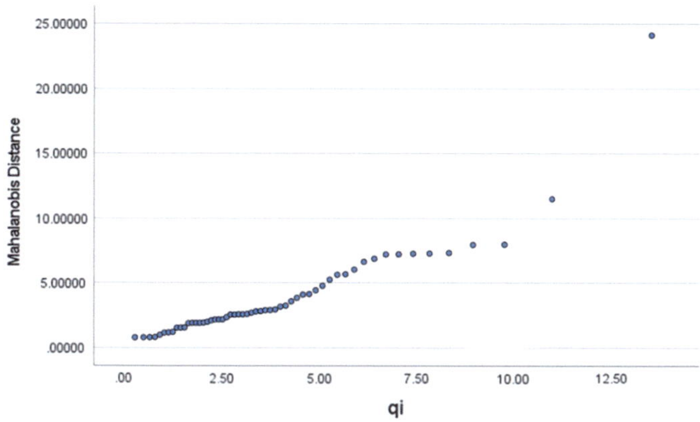

Fig. 2 Q–Q plots

Table 2 Mahalanobis distance correlation results

		Mahalanobis distance	qi
Mahalanobis distance	Pearson correlation	1	0.921
	P-value		< 0.001
	N	56	56

3. Multicollinearity Assumptions

The multicollinearity assumption can be detected by the VIF value. The following will display the VIF value of each variable on a ratio or numerical scale used in this study. In this research, according to the findings presented in Table 3, the variance inflation factor (VIF) values for each variable indicated no significant interrelation, as all VIF values were below 10. Consequently, this study satisfies the multicollinearity assumption.

Table 3 VIF results

Variable	VIF
X_1	1.100
X_2	1.044
X_3	1.052
X_4	1.148

Table 4 Cluster validity index

Number of cluster	Number of iteration	Time	Silhouette	Xie-Beni index
2	**117**	**0.8**	**0.349**	**0.776**
	86	0.5	0.311	1.024

3.3 *Clustering Using the Gath-Geva Algorithm*

The factors that influence customer decisions in choosing sea transportation modes will be grouped into two, namely internal factors and external factors using the Gath-Geva algorithm with the number of clusters used are two as shown in Table 4. Table 4 provides information that using a number of clusters of 2, namely to classify variables into two groups, namely internal factors and external factors where the iteration with the highest silhouette validity value and the lowest Xie-Beni is at the 117th iteration with a time of 0.8 s.

The number of members of each cluster is as shown in Table 5. The information obtained from Table 5 is that the majority choose to use sea transportation services for personal reasons or internal factors themselves, in this case the variable percentage of service expectations (X_2). The details of the members of each cluster based on the grouping results with the Gath-Geva clustering algorithm can be seen in Fig. 3. Based on Fig. 3, it is known that the clustering results can explain the cluster variability of 62.44%, where this percentage is high enough so that the clustering results are good enough.

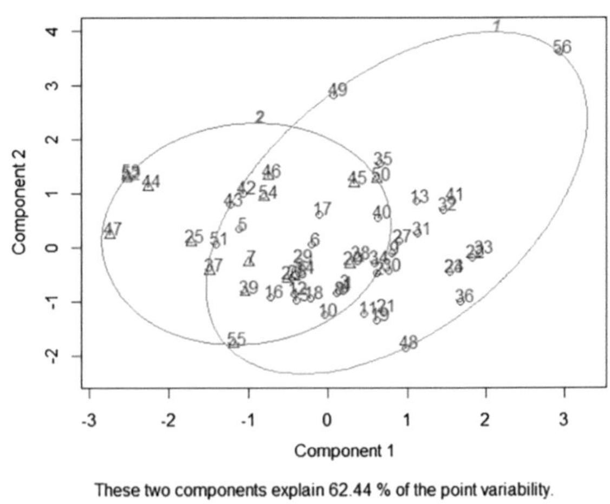

Fig. 3 Members of each cluster

Table 5 Number of members of each cluster

Cluster	Number of member	Factor
1	39	Internal
2	17	External

4 Conclusion

The analysis conducted reveals that a significant portion of sea transportation service users, 46.43%, utilize services at the port of Java island, while the remaining 53.57% use services outside of Java island. Analysis using the Gath-Geva clustering method in this study has fulfilled all the necessary assumptions, namely multivariate normal assumptions, data adequacy assumptions and multicollinearity assumptions and it is known that the majority of people who use sea travel services are based on internal factors or factors that come from within themselves, namely service expectations (X2). Thus, if a company providing sea travel services wants to be in great demand by the public, the services provided must be in accordance with the wishes of the customer by conducting market research on sea transportation to the community, especially regarding the expected facilities and services. Clustering using the Gath-Geva algorithm can produce a classification accuracy of 62.44%, which is a good percentage. For further research, it is expected to be able to explore more deeply and more about these internal and external factors so that more factors can be evaluated for the policies of shipping service providers.

Acknowledgements Gratitude is extended to the Department of Marine Transportation Engineering at the Institut Teknologi Sepuluh Nopember, Surabaya, along with the Directorate General of Higher Education, Research, and Technology, under the Ministry of Education, Culture, Research, and Technology of the Republic of Indonesia, for their support in this research.

References

1. Kementrian Perhubungan Republik Indonesia: Pembangunan Pelabuhan untuk merajut konektivitas transportasi di Sulawesi. https://dephub.go.id/post/read/pembangunan-pelabuhan-untuk-merajut-konektivitas-transportasi-di-sulawesi (2023). Acceses 13 Sept 2023.
2. R. Syoer, M. Mashuri, Analisis kelompok dengan algoritma fuzzy c-means dan gath-geva clustering. IndoMS J. Statistics **2**(2), (2014). Acceses 11 Sept 2023
3. Y. Puspitasari, W. Syaifudin, R.D. Nofyandi, Pemetaan daerah rawan kecelakaan menggunakan metode fuzzy c-means, 90–95 (2019). Acceses 10 Sept 2023
4. M. Li, M.K. Ng, Y. Cheung, Agglomerative fuzzy k-means clustering algorithm with selection of number of clusters. IEEE Trans. Knowl. Data Eng. 20 (2008). Acceses 13 Sept 2023
5. E. Rusli, Pengujian normal multivariat dan homoskedastisitas matriks varians-kovarians pada prestasi belajar dan kredit lulus mahasiswa unm angkatan 2014. (Ilmu Pendidikan UNM, Malang, 2014). Acceses 16 Sept 2023
6. F.B. Opong, S.Y. Agbedra, Assessing univariate and multivariate normality, a guide for non-statisticians. Mathematical Theory and Modeling **6**(2), 27 (2016). Acceses 11 Sept 2023
7. R. Johnson, D. Wichern, *Applied Multivariate Statistical Analysis*, 6th edn. (Prentice Hall International, New Jersey, 2007). Acceses 10 Sept 2023
8. J. Hair, W. Black, B. Babin, R. Anderson, *Multivariate Data Analysis*, 7th edn. (Pearson, 2009). Acceses 12 Sept 2023
9. I. Ghozali, *Aplikasi Analisis Multivariate Dengan Program SPSS* (Badan Penerbit UNDIP, Semarang, 2017). Acceses 18 Sept 2023

10. N. Shrestha, Detecting multicollinearity in regression analysis. Am. J. Appl. Math. Statist. **8**(2), 39 (2020). Acceses 13 Sept 2023
11. A.N. Azizah, T. Widiharih, A.R. Hakim, Kernel k-means clustering untuk pengelompokkan sungai di kota semarang berdasarkan faktor pencemaran air. J. Gaussian **11**(2), 232 (2022). Acceses 17 Sept 2023

Electromyography Finger Movement Classification

Saurabh Kumar Srivastava, Shilpi Singh, Saurabh Sambhav, Aman Kumar Singh, and Mukund Pratap Singh

Abstract Finger movement identification is a crucial method for advanced interfacing, with applications ranging from innovative HCI devices to medical prosthetics. Current approaches, such as computer vision and vibration-based detection, face limitations like restricted hand movement and environmental sensitivity. This study focuses on rectifying shortcomings in existing EMG-based hand and finger gesture recognition systems, emphasizing individual finger gestures. The proposed solution involves EMG-based classifiers with fixed electrode placement and machine learning techniques. Despite existing models like ANN and SVM, challenges persist. The paper advocates for a CNN, achieving an impressive 83.2% average classification accuracy, outperforming traditional methods. This research significantly enhances EMG-based gesture recognition, particularly for individual finger movements.

Keywords Gesture recognition system · SVM · ANN · CNN · Healthcare

S. K. Srivastava · M. P. Singh
School of Computer Science Engineering and Technology, Bennett University, Greater Noida, India
e-mail: phd.jiit@gmail.com

M. P. Singh
e-mail: a.mps152@gmail.com

S. Singh (✉) · S. Sambhav · A. K. Singh
Amity University Patna, Patna, India
e-mail: ssingh3@ptn.amity.edu

S. Sambhav
e-mail: ssambhav@ptn.amity.edu

A. K. Singh
e-mail: rajputaksingh78@gmail.com

A. Ismail et al. (eds.), *Advancements in Industry and Technology*,
SpringerBriefs in Applied Sciences and Technology,
https://doi.org/10.1007/978-3-031-97931-6_12

1 Introduction

The advent of finger movement identification marks a transformative milestone in human–computer interfacing (HCI), offering a myriad of applications that span from revolutionary HCI devices to groundbreaking developments in medical prosthetics. This innovative method stands as a beacon of progress, holding the potential to redefine the way individuals interact with technology and addressing critical needs in the medical field. Current research endeavors in this domain explore diverse avenues, from computer vision techniques to detecting finger motion through mechanical vibrations on the skin surface. However, these methods exhibit inherent limitations, notably in terms of constraining the degree of hand movement from a specific optimum position and susceptibility to environmental factors. As technology advances, it becomes imperative to overcome these constraints and propel the field toward more sophisticated solutions.

In this context, the research at hand assumes a crucial role by delving into the limitations of existing electromyogram (EMG)-based hand and finger gesture recognition systems. Notably, the majority of current systems predominantly focus on recognizing gestures involving the wrist and the entire hand, neglecting the nuanced intricacies of individual finger (IF) gestures. Recognizing the significance of individual finger movements, the study aims to rectify these limitations and propel the field toward a more comprehensive and precise understanding of hand gestures.

The proposed approach involves the utilization of EMG-based hand/finger gesture classifiers that leverage fixed electrode placement and innovative machine learning techniques.

While various classification models such as artificial neural networks (ANN), artificial neural networks have been proposed for hand and arm movement recognition, persisting challenges include inaccurate classification, insufficient generalization ability, and weak robustness. To overcome these hurdles, this paper advocates for the integration of a deep learning model, specifically artificial neural networks (CNN), to enhance the classification accuracy and robustness of the system. The study involves a dataset comprising eight classes of hand gestures with participants seated and their arms supported, aiming to provide a comprehensive evaluation of the proposed deep learning approach. Through this research, the goal is to contribute significantly to the evolution of EMG-based hand and finger gesture recognition, particularly by addressing the nuances of individual finger gestures. By doing so, the study aims to usher in a new era of more accurate, versatile, and applicable human–computer interfacing technologies, with potential implications for diverse fields, including HCI devices and advanced prosthetics.

2 Related Work

Based on findings from [1], the author reports that utilizing K-nearest neighbor yielded an average classification accuracy of 80.60%, whereas employing linear discriminant analysis (LDA) resulted in an average accuracy of 81.84%. The analysis reveals that LDA exhibited the highest level of accuracy in classification. The paper [2] introduces a technique for identifying finger motions by utilizing the wavelet transform of multi-channel electromyography signals. The initial phase involves analyzing surface EMG signals through multi-resolution wavelet transform.

Next, the process involves inputting the extracted feature values into an artificial neural network to recognize the finger motion. As indicated by [3], EMG signals contain valuable information about the nervous system. This study elucidates the diverse range of EMG signal analysis techniques, facilitating their appropriate application in clinical diagnosis, biomedical research, hardware implementations, and end applications.

As stated in [4], convolutional neural networks (CNNs) represent a relatively new artificial neural network architecture. Due to its utilization in image and speech recognition, resulting in enhanced test outcomes, this algorithm has gained widespread adoption and application. CNNs dominate the realm of computer image recognition, yet due to continual innovation, they have found applications in video analysis, drug analysis, natural language processing, and various other domains. In this sequence of research papers [5–8], researchers have devised a method utilizing wavelet transform to classify single motor unit potentials and to break down EMG signals into their individual SMU potentials.

This approach is specifically designed for disentangling multi-unit EMG signals marked by substantial waveform overlap, rendering it inappropriate for real-time analysis. The extent of human involvement typically corresponds with the quantity of active SMU potentials detected in EMG signals [9]. Nonetheless, neither morphological nor frequency domain techniques surpass time domain features. Although principal component analysis improves results marginally, it incurs a time cost.

Furthermore, paper [10] underscores that surface electromyography recordings of skeletal trunk muscles often encounter spike-shaped artifacts originating from electrical heart activity, which are commonly detected by electrocardiography and frequently present in surface electromyography signals captured near the heart. However, this study does not address the relatively higher fluctuations of the ECG signal during surface electromyography recording.

The objective of research [11] is to pinpoint the optimal features to achieve a high rate of motion classification for controlling an artificial hand. Nineteen EMG signal features are assessed, with recommendations including the combination of wavelet transform with other signal processing techniques. An evaluation is conducted from three perspectives: (i) motion classification, (ii) noise tolerance, and (iii) computational complexity. The energy of wavelet coefficients of EMG signals across nine scales and the cepstral coefficients emerge as the most effective features in these evaluations.

In this paper [12], a novel hybrid CNN model is presented for classifying hand gestures using sEMG data. The model integrates elements from convolutional neural networks architectures, incorporating two parallel inception modules initially. The first module processes data from four electrodes, while the second module handles the remaining electrodes, aligning with the typical behavior of slow-fusion architecture that processes split data in parallel. Subsequently, the outputs of the Inception models are fused throughout the network. The input feature maps are constructed in 3D using short-time Fourier transform (STFT) applied to raw data from all electrodes across time-series. Experimental results demonstrate that the proposed method achieves an accuracy of 83.97% with a standard deviation of 0.027. Moreover, it is noted that the introduced model outperforms both slow-fusion models when used individually.

3 Proposed Framework

A. Data Collection

The raw surface electromyography signal exhibits a broad frequency range, extending up to 6500 Hz, with the primary spectral power concentrated from the range of 20–150 Hz. To enhance signal quality, frequencies below 20 Hz are eliminated using a high-pass filter. Additionally, frequencies above 450 Hz are filtered out using another filter (butterworth), following the recommendation for optimal signal filtering. Post-filtering, the absolute value of the EMG signal is calculated to prevent the cancellation of positive and negative values during mean calculation. This step reduces the frequency content of the EMG signal, thereby minimizing memory storage requirements and facilitating interpretation and activity onset detection.

A total of 6823 samples were collected, with 70% designated for training and the remaining 30% for testing purposes. The training set underwent 20000 epochs during experimentation, with an average prediction time per sample of 3.9 milliseconds.

The research methodology outlined in this paper encompasses three main steps: data collection, data pre-processing, and classification.

B. Data Pre-processing

The EMG potential undergoes oscillations between positive and negative values, resulting in cancellations when computing various features like mean and average. Baseline EMG values often exhibit an offset, especially during contraction. Following this, EMG signals are subjected to filtering, with power line noise removed using a built-in notch filter. The raw surface EMG signal encompasses a wide frequency range from 6 to 500 Hz, with most of the spectral power concentrated between 20 and 150 Hz. To refine the signal, frequencies below 20 Hz are eliminated using a high-pass filter applied to the mean-corrected EMG signal, while frequencies above 450 Hz are filtered out using Butterworth filters. After filtering, the absolute value of the EMG signal is calculated to prevent the cancellation of positive and negative values during mean calculation. This action reduces the frequency content of the EMG signal, making it easier to interpret and detect activity onset while also minimizing memory storage requirements.

C. Signal Classification

A grand total of 6823 samples were gathered, with 70% designated for training and the remaining 30% set aside for testing. Throughout the experimentation process, the training set underwent 20,000 epochs, with an average prediction time per sample clocking in at 3.9 ms. The dataset encompassed an array of finger gestures, encompassing flexion of the index, middle, ring, and little fingers, thumb movement, resting position, and the victory gesture. The analysis was conducted with a dual focus to evaluate the efficacy of pre-processing. A comprehensive exploration involving various algorithms such as pre-processing was carried out to gauge their effectiveness. Pre-processing emerged as a potent deep learning methodology over the last two decades, surpassing earlier models. With advancements in architecture and the utilization of graphics processing units (GPUs) for parallel computing, CNNs have become notably more efficient. This paper underscores the potential of CNNs as a versatile tool in the realms of bio-robotics and rehabilitation.

4 Results and Discussion

We conducted an analysis of EMG signals obtained from volunteers, focusing on feature extraction and the application and comparison of SVM, ANN, K-NN, and CNN classifiers. Furthermore, we employed these classifiers in the classification of finger movements.

We conducted an in-depth analysis of electromyography signals obtained from volunteers, involving comprehensive steps such as feature extraction and the application and comparison of various classifiers, including graphics processing units, graphics processing units.

The primary objective of this investigation was to establish a robust classification system for all finger movements. In the signal classification process, we initially acquired electromyography signals using gold-plated surface electrodes. Subsequently, we fed this data into our CNN model, which initially partitioned the data into a 70–30 ratio for subsequent testing and cross-validation. The training and testing phases involved the utilization of backpropagation. Through the steps and the implementation of the CNN algorithm, we achieved superior accuracy compared to other algorithms, emphasizing the effectiveness of our approach in achieving precise and reliable classification of finger movements.

Figures 1 and 2 show the accuracy and loss graphs over the training. We trained the 1D CNN model over 40000 epochs but after 20000 epochs the loss and accuracy became constant of about 0.8 accuracy and 1.2 loss. We split the dataset into 30–70%. 70% of the data was used to train the model and 30% was used to test. We used 80–20 ratio, 90–10, and several other ratios but we saw the most optimal result in 70–30 splits. The error was calculated after each epoch and was backpropagated to each of the layers and after tuning up and down the learning rate we found the most optimal learning rate so that the model does not stick in minima's and do not overshoots was found to be 0.0001. The sudden drop in accuracy is represented when the model was stuck in the local minima.

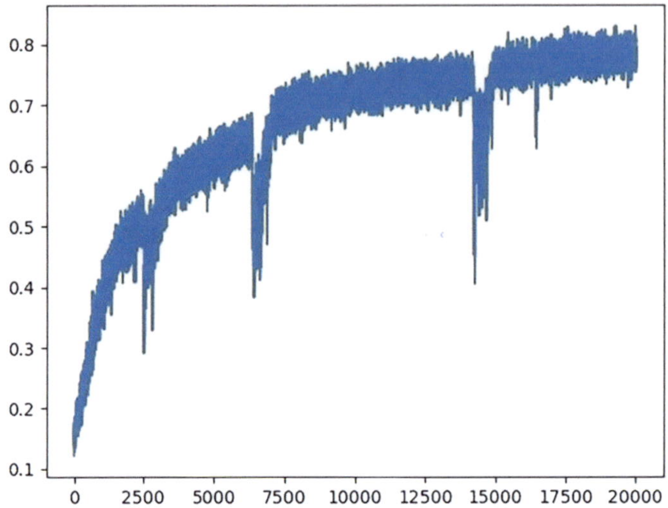

Fig. 1 Accuracy graph (Y-axis is for accuracy and X-axis is for number of epochs)

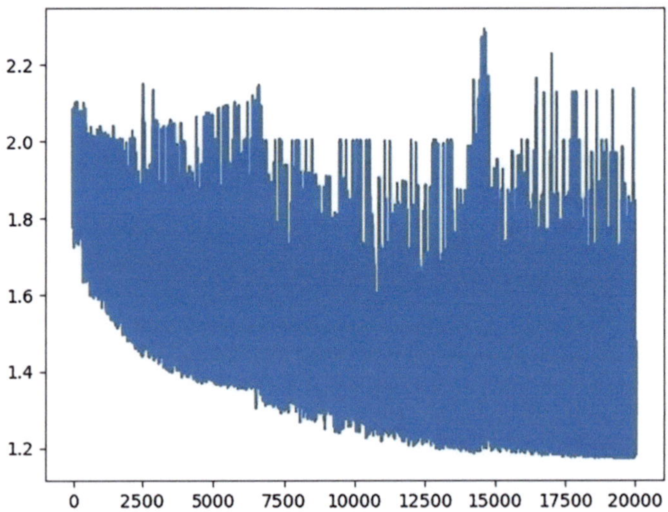

Fig. 2 Loss graph (Y-axis is for loss and X-axis is for number of epochs)

5 Conclusion

In conclusion, this research successfully addressed limitations in current electromyogram (EMG)-based hand and finger gesture recognition systems, particularly those focusing on individual finger gestures. The study introduced a novel approach, leveraging a convolutional neural network, which demonstrated a remarkable average classification accuracy of 83.2%. This signifies a substantial improvement over traditional methods and points to the viability of deep learning models in enhancing the accuracy and robustness of hand gesture recognition systems. The findings not only contribute to the scientific understanding of EMG-based gesture recognition but also hold practical implications for the development of more sophisticated and precise human–computer interfacing technologies. The successful implementation of the proposed CNN model paves the way for future advancements in the field, fostering innovation in diverse applications such as HCI devices and advanced prosthetics.

References

1. S. A. Fandaklı, Ö. Aydemir, A fast and highly accurate EMG signal classification approach for multifunctional prosthetic fingers control, in *40th International Conference on Telecommunications and Signal Processing (TSP)* (IEEE, 2017), pp. 395–398
2. M. W. Jiang, R. C. Wang, J. Z. Wang, D. W. Jin, A method of recognizing finger motion using wavelet transform of surface EMG signal, in *IEEE Engineering in Medicine and Biology 27th Annual Conference* (IEEE, 2006), pp. 2672–2674
3. M.B.I. Reaz, M.S. Hussain, F. Mohd-Yasin, Techniques of EMG signal analysis: detection, processing, classification, and applications. Biol. Proced. Online **8**(1), 11–35 (2006)
4. E.P. Ijjina, K.M. Chalavadi, Human action recognition using genetic algorithms and convolutional neural networks. Pattern Recogn. **59**, 199–212 (2016)
5. J. Fang, G. C. Agarwal, B. T. Shahani, Decomposition of EMG signal by wavelet spectrum matching, in *Proceedings of the 19th Annual International Conference of the IEEE Engineering in Medicine and Biology Society. 'Magnificent Milestones and Emerging Opportunities in Medical Engineering'(Cat. No. 97CH36136)*, vol. 3. (IEEE, 1997), pp. 1253–1256. https://doi.org/10.1109/IEMBS.1997.756598.
6. R.S. LeFever, C.J. De Luca, A procedure for decomposing the myoelectric signal into its constituent action potential-part I: technique, theory, and implementation. IEEE Trans. Biomed. Eng. **3**, 149–157 (1982)
7. R.S. LeFever, A.P. Xenakis, C.J. De Luca, A procedure for decomposing the myoelectric signal into its constituent action potentials-part II: execution and test for accuracy. IEEE Trans. Biomed. Eng. **3**, 158–164 (1982)
8. S. Sambhav, S. Singh, Low-power IoT architecture, challenges, and future aspects, in *International Conference on Micro-Electronics and Telecommunication Engineering* (Springer Nature Singapore, Singapore, 2021), pp. 553–560
9. M.G. Jahromi, H. Parsaei, A. Zamani, D.W. Stashuk, Cross comparison of motor unit potential features used in EMG signal decomposition. IEEE Trans. Neural Syst. Rehabil. Eng. **26**(5), 1017–1025 (2018)
10. N. Miljković, N. Popović, O. Djordjević, L. Konstantinović, T.B. Šekara, ECG artifact cancellation in surface EMG signals by fractional order calculus application. Comput. Methods Programs Biomed. **140**, 259–264 (2017)

11. R. Boostani, M.H. Moradi, Evaluation of the forearm EMG signal features for the control of a prosthetic hand. Physiol. Meas. **24**(2), 309 (2003)
12. S. Sharma, G. Kumar, S. Kumar, D. Mohapatra, Techniques for feature extraction from EMG signal. Int. J. Adv. Res. Comput. Sci. Softw. Eng. 2, no. 1 (2012).

Awareness of Maritime Piracy in the Sulu Sea Among Nautical Students

Nur Farisha Mohamad Fauzi, Mohammed Ismail Russtam Suhrab, and Ahmad Faizal Ahmad Fuad

Abstract Piracy and maritime terrorism are global threats to maritime security, with a long history of occurrence across the world. In particular, the Sulu Sea region has been impacted by piracy activities, including the Abu Sayyaf attack on the Sabah area. This attack has contributed to increased piracy cases in the Sulu Sea. The objective of this research is to provide nautical students with knowledge and awareness of the requirements that must be met in the event of a piracy attack by identifying the factors that lead to pirate attacks on vessels in the Sulu Sea and determining the type of ships that are most commonly targeted. The research method includes a literature review and questionnaire, and the data will be analyzed using descriptive analysis, SPSS, and judgment. Although piracy attacks have decreased in recent years, the possibility of future attacks cannot be ruled out, as the Abu Sayyaf Group remains at large. The findings of this research will contribute to reducing piracy attacks in the future near the Sulu Sea. This research will provide valuable information to nautical students and equip them with the knowledge and requirements needed to respond to piracy attacks.

Keywords Piracy · Sulu Sea · Piracy aimed ship type · Practice to counter piracy

1 Introduction

Piracy and maritime terrorism are longstanding problems in maritime law and navigation and continue to threaten maritime security [1]. Efforts to combat these crimes have been ongoing for many years, with major players in maritime policy discussing

N. F. M. Fauzi · M. I. R. Suhrab (✉) · A. F. A. Fuad
Program of Nautical Science and Maritime Transportation, Faculty of Maritime Studies, Universiti Malaysia Terengganu, Terengganu, Malaysia
e-mail: m.ismail@umt.edu.my

N. F. M. Fauzi
e-mail: s54828@ocean.umt.edu.my

A. F. A. Fuad
e-mail: afaf@umt.edu.my

A. Ismail et al. (eds.), *Advancements in Industry and Technology*,
SpringerBriefs in Applied Sciences and Technology,
https://doi.org/10.1007/978-3-031-97931-6_13

possible countermeasures. There is a lot of terrorism going on, and it has been a problem for many countries. One of the countries that has been affected is Malaysia. Malaysia is on the island of Borneo, which is close to three other countries—Indonesia, the Philippines, and Brunei. There are a lot of threats to Malaysia's survival, especially in the waters near the Strait of Malacca. This is where pirates have been attacking ships.

Today, the waters of the Sulu Sea are still known as a piracy hotspot in Southeast Asia [2, 3]. This reputation is not based on the number of attacks in international statistics published by organizations such as the International Maritime Bureau (IMB). Indeed, attacks on larger vessels rarely take place in the waters between Sabah and the southern Philippines, even though petty thefts from larger vessels occur in some ports found in the area. Most pirate attacks in these waters target small vessels such as fishing boats, transport vessels, and small passenger ships and remain unreported. Moreover, attacks on vessels are only one type of piratical activity prevalent in this area, with raids on towns and offshore businesses conducted by armed men traveling in boats a frequent security concern in Sabah. The piracy issue in the Sulu Sea is the Abu Sayyaf attack on their claim that part of the Sabah area belongs to them.

Problem Statement

The Abu Sayyaf attack in the Sulu Sea was a pirate incident in which they claimed ownership of a part of Sabah's territory [2]. The high frequency of piracy attacks in the Sulu Sea area has made the ocean between Sabah and the southern Philippines one of Malaysia's piracy hotspots. The Abu Sayyaf attacks have resulted in an uptick in pirate cases in the Sulu Sea. (2018 Annual Report on Piracy and Armed Robbery Against Ships in Asia). Even the incident from the Abu Sayyaf group was decreasing year to year. However, this does not rule out the possibility of another future piracy attempt in the Sulu Sea, as the Abu Sayyaf Group has yet to be apprehended. So, this research will supply fresh information and help nautical students gain knowledge and what they need to do about requirements that need to be supplied when piracy attacks happen. Figure 1 shows the map of the Sulu Sea.

Research Objective

The aim of this research is to:

I. Find the factors that lead pirates to attack a vessel or ship in the Sulu Sea.
II. Decide the majority type of the ship that was attacked.
III. Collect suggestions and advice from nautical students about requirements that must be supplied when piracy attacks happen for upcoming new seafarers.

Piracy Attack

Piracy and maritime terrorism pose a growing threat to maritime security [4]. Hotspots for piracy include the Gulf of Aden, Malacca Strait, New Guinea, and the Sulu Sea, where vessels face frequent pirate attacks. In 2016, there were about

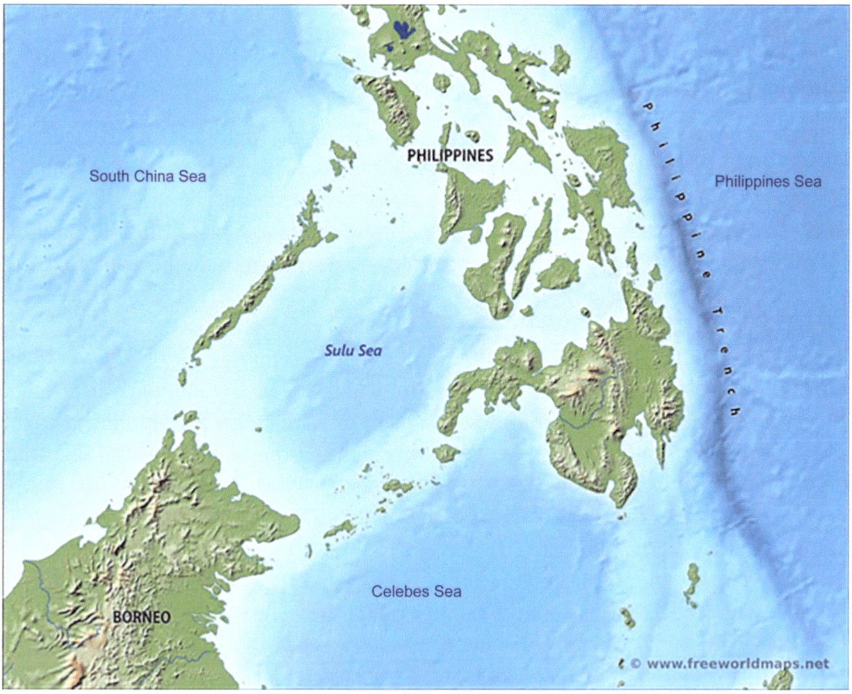

Fig. 1 Sample 1. Source https://tempo.com.ph/2020/10/22/enhanced-monitoring-needed-to-preserve-tubbataha-reef-in-sulu-sea/23map-of-sulu-sea/

20 offshore assaults, leading to numerous casualties and abductions. This situation raises concerns about the Sulu Sea becoming a "New Somalia." Border countries (Indonesia, Malaysia, and the Philippines) have taken cooperative measures to combat piracy, resulting in fewer incidents but still significant challenges in the region [5–7].

2 Methodology

When answering the target, collect data, such as observations from literature reviews and questionnaires. This section of the research shows the methodology, techniques, and data collection methods used. This section also describes the methods and procedures used to collect and analyze data to determine the research objective, that is, to collect knowledge and spread awareness among nautical students about requirements that need to be provided when piracy attacks happen by identifying the factors that lead pirates to attack a vessel or ship in the Sulu Sea and determine the type of the ship that was attacked.

Research Design

Study on Literature Review

First, the sources of piracy were discovered by studying literature reviews, such as review articles, books, and online sources. Reviews of the literature are primarily referred to as secondary data. This study aims to collect and identify information from reference materials. These data come more from sources written by authors and published in different forms, for example, books, journals, and the internet.

Data Collection

The first objective aims to identify the factors that lead pirates to attack a vessel or ship in the Sulu Sea. These findings were compiled based on five comparison articles/journals to evaluate each piece of information about factors that lead pirates to attack a vessel or ship in the Sulu Sea [8]. Not only that but the findings of the literature study and descriptive analysis were also obtained.

The second objective is to determine the majority type of the attacked ship. ICC International Maritime Bureau (2018) collected these findings. Also, this collected data has been supported through an online questionnaire given to nautical students at the University Malaysia Terengganu to collect their knowledge and suggestions about these findings' objectives. Globally, many types of vessels were attacked. Still, for this research, only most ships were selected to determine the second objective, and these collected ship types are listed in Fig. 2 (ICC International Maritime Bureau, 2018). Figure 2 shows that the number of attacks in the past seven years was still alarming. In the latest year, 2018, bulk carriers, tanker chemical products, and container vessels were the main targets for attacks, with several attacks of 59, 50, and 18, respectively [9]. According to the ICC International Maritime Bureau (2018), Malaysia was recorded as the second-highest country in the number of actual and attempted attacks from 2013 to 2018, with 71 cases, after Indonesia (342 cases) [10].

The third objective is to learn about the piracy attack and incident in the Sulu Sea from nautical students. This research was done with a specially developed questionnaire. Concerning data collection, the questionnaire was made in Google form and developed as an online questionnaire to organize an online survey. Flexibility, accessibility, and the potential to enter broad samples are the advantages of the online survey, governing among the main strengths of the online survey. The online questionnaire, distributed among nautical students at the University Malaysia Terengganu, was set up to obtain data for this research. This third research collects suggestions and advice from nautical students about requirements that need to be provided when piracy attacks happen for upcoming new seafarers.

Development of Questionnaire

The questionnaires are used to accomplish the second and third objectives. The questionnaire consisted of two types of questions. The first type of question for the second objective allows the respondent to fill in the answers in the space below the

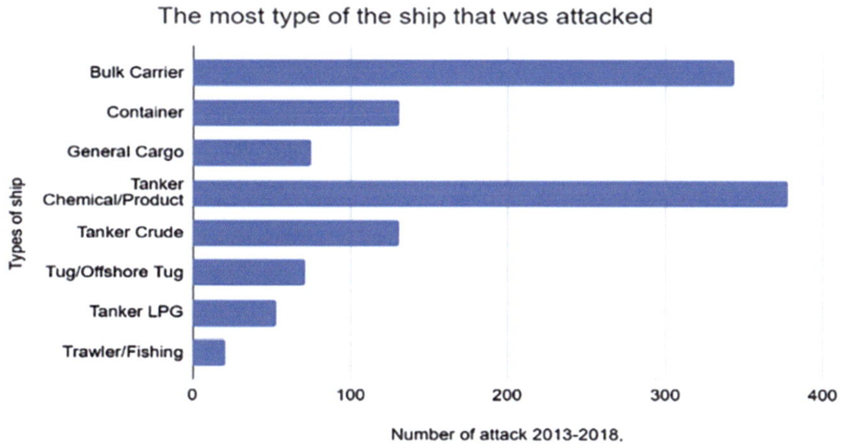

Fig. 2 Most type of ship that was attacked from 2013 to 2018. *Source* ICC International Maritime Bureau (2018)

question. The second type of question that was used to accomplish the third objective is a five-point rating scale question. The weights assigned to each answer choice are shown in parentheses: Strongly Agree, Agree, Neutral, Disagree, and Strongly Disagree. The Likert scale was used to set up the questionnaire. A Likert scale is a psychometric scale commonly involved in research that employs questionnaires. It is the most widely used approach to scaling responses in survey research. The term is often used interchangeably with rating scales or, more accurately, the Likert-type scale, even though the two are not synonymous. The scale is named after its inventor, psychologist Renzi's Likert [11].

Data Analysis

Descriptive analyses are the first step in data analysis. The descriptive analysis aims to determine which type of ship was attacked most often. It is only nominal data and sometimes ordinal data; it is intended to shape this data for efficient use of data. The easiest and fastest way to examine the collected data is by using (frequency) tables and bar charts [12]. Following the survey, the nautical students showed that they also have experience in answering these questions for this questionnaire. It is about reviewing and refining ideas, examining consensus and disagreement, and examining strengths and weaknesses. Data analysis can be performed using a suitable computer program like Microsoft Excel. Finally, the chart can be plotted using previously collected data sufficient to determine measures of central tendency in the statistics, such as the mean and maximum value of the chart. After completing these data collection and analysis steps, discussions can be made using the results obtained to make recommendations to improve a vital issue of this study.

3 Results and Discussion

Demographic of Respondent

After the questionnaires have been distributed to the respondents, consisting of first-year until fourth-year nautical students studying at the University Malaysia Terengganu, as shown in the methodology, the respondents responded according to their respective batches to the survey report. The following information is obtained from all the respondents. A total of 25 respondents were third-year nautical students, comprising 37.9% of the respondents, while another 62.1% were first, second, and fourth-year nautical students, with 20 respondents from the fourth year, 11 from the first year, and 10 from the second year. All his 66 participants had basic knowledge about the question that was given.

Respondent's Knowledge of the Topic of Research

This objective was supported through an online questionnaire given to nautical students at the University of Malaysia Terengganu. The first section of the questionnaire focused on the level of respondents' knowledge based on their study related to the research topic to determine the majority type of ship that was attacked. The question was why these types of ships are easy to attack. From the question that was asked, most nautical students agreed that the majority of types of vessels are easy to attack. Figure 3 shows that the fourth- and third-year nautical students have more knowledge about piracy issues since they were the seniors among nautical students.

The second section of the questionnaire also focuses on respondents' knowledge based on their study related to the research topic: to collect suggestions and advice from nautical students about requirements that must be provided when piracy attacks

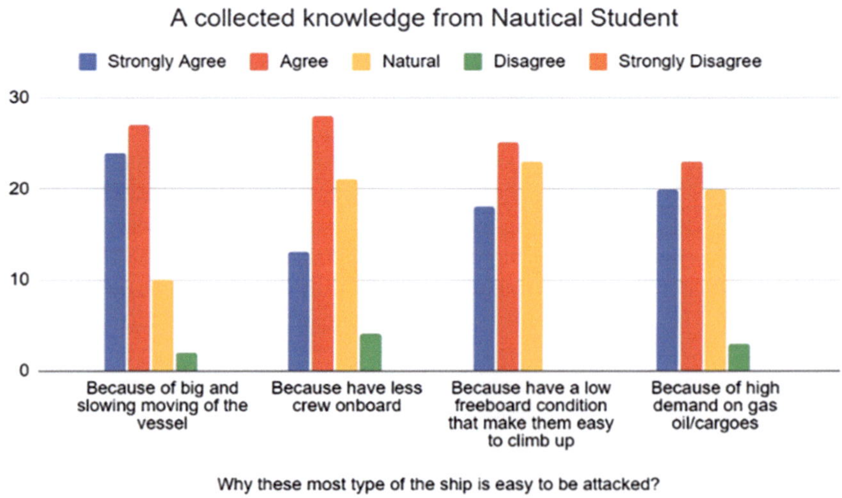

Fig. 3 Graph shows collected data for answering the objective

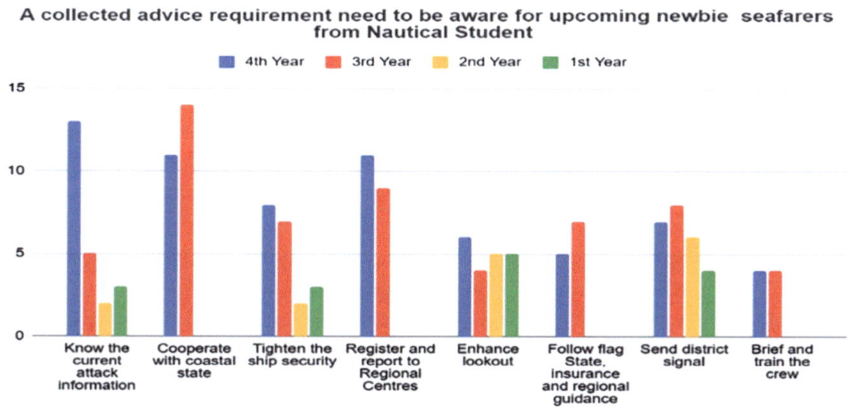

Fig. 4 Graph shows collected data for answering objective 3

happen for upcoming new seafarers. This objective was collected through an online questionnaire from nautical students at the University of Malaysia Terengganu. The question asked is what action and requirement of upcoming new seafarers must be provided when piracy attacks happen. From the question that was asked, most Nautical senior students know what action they need to be aware of if they are in that situation. Figure 4 shows that the fourth- and third-year nautical students have more knowledge about requirements that need to be provided when piracy attacks happen for upcoming new seafarers since they were the seniors among Nautical students.

Result Objective Archive

Objective 1 identifies the factors that lead pirates to attack a vessel or ship in the Sulu Sea. The results were obtained by comparing five articles and journals. The result states that a desire to gain wealth that leads to maritime terrorism in the Sulu Sea has the most mentioned data from five articles. Also, followed by the maritime traffic mentioned in four articles, lack of security cooperation was mentioned in three articles, and two articles that were collected state that the factor that leads to piracy is because certain communities have been known as pirate societies. For aim 2, most types of ships attacked by piracy were found by a reasonable answer from a literature review and article. Also, from knowledge of nautical students from UMT, most agree with these reasons for the attack. Last, objective 3 was achieved by collecting through an online questionnaire that was given to nautical students at the University Malaysia Terengganu. The data shows that the fourth year and third year of nautical students have more knowledge about requirements that need to be provided when piracy attacks happen for upcoming new seafarers since they were the seniors among nautical students. From the question that was asked, most Nautical senior students know what action they need to be aware of if they are in that situation.

4 Conclusion

The analysis of previous years' statistics has indicated that piracy and maritime terrorism activities persist, involving a variety of violent acts, vessels, and weapons. This paper examines piracy from multiple perspectives, investigating the level of knowledge and awareness among maritime students regarding the measures that must be taken in the event of a piracy attack. The study considers the factors that motivate pirates to target a particular vessel and the types of ships that are most frequently targeted. A literature review and questionnaire are employed to gather data, which is subsequently analyzed using descriptive analysis, facilitated by SPSS. Conclusions are then drawn through the application of judgment.

Piracy is a critical issue that requires the utmost attention of maritime students and the Malaysian government. To prevent attacks, it is imperative that students are aware of the necessary requirements and that the government implements appropriate security measures. The Malaysian government should take steps to ensure maritime security in the Sulu Sea by strengthening cooperation with other ASEAN members in counterterrorism efforts. The involvement and cooperation of various agencies, including the Malaysian Armed Forces, Royal Malaysian Police, and Malaysian Maritime Enforcement Agency, are crucial in addressing piracy and terrorism issues. Despite the establishment of ESSCOM, kidnappings of Malaysian citizens in Sabah by ASG groups continue to occur, highlighting the need for increased cooperation with ASEAN in resolving transnational threats. Cooperation with ASEAN will provide additional strength to Malaysia in resolving these issues. The global consequences of piracy require countries to work toward common goals in security, including counterterrorism, human trafficking, and drug smuggling. The advancements in communication and transportation technologies have made it increasingly challenging to control state borders from outside threats. It is imperative that the government enhances and strengthens the coordination between relevant agencies to ensure maritime security in the Sulu Sea.

References

1. S. E. Amirell, *The Return of Piracy: Decolonization and International Relations in a Maritime Border Region* (the Sulu Sea, 2005), pp. 1959–63
2. C. Liss, *Assessing Contemporary Maritime Piracy in Southeast Asia: Trends, Hotspots, and Responses* (Peace Research Institute Frankfurt, Frankfurt, 2014)
3. Liss, C.: Piracy and maritime violence in the waters between Sabah and the southern Philippines, in *Piracy in Southeast Asia* (Routledge, 2016), pp. 165–181
4. Ariani, S.: The Implementation of Trilateral Cooperation Arrangement Between Indonesia, Malaysia, And the Philippines: Increasing of Security Cooperation to Reduce Maritime Piracy at Sulu Sea Dissertation, President University, 2018
5. N.H.A. Karim, N.H.M. Salleh, M. Harun, Maritime terrorisms and navigational security in the Sulu sea. Int. J. E-Navig. Marit. Econ. **12**, 42–54 (2019)
6. M. F. Chew, *Piracy, Maritime Terrorism, and Regional Interests* (2005)

7. S. Febrica, Securing the Sulu-Sulawesi seas from maritime terrorism: troublesome cooperation? Perspect. Terror. **8**(3), 64–83 (2014)
8. J.V. Hastings, The return of sophisticated maritime piracy to Southeast Asia. Pac. Aff. **93**(1), 5–30 (2020)
9. I. Storey, *Addressing the Persistent Problem of Piracy and Sea Robbery in Southeast Asia* (ISEAS-Yusof Ishak Institute, 2016)
10. C.M. Nguyen, T.Q. Le, Impact of piracy on maritime transport and technical solutions for prevention. Int. J. Civ. Eng. Technol. **10**(01), 958 (2019)
11. N.H.A. Karim, N.H.M. Salleh, M. Harun, Maritime terrorisms and navigational security in Sulu sea. Int. J. E-Navig. Marit. Econ. **12**, 42–54 (2019)
12. S. Moreels, The insurability of maritime terrorism. Unpublished thesis, Gent University, 2016. https://lib.ugent.be/fulltxt/RUG01/002/272/401/RUG01-002272401_2016_0001_AC

Factors Contributing Toward Effective Maritime Simulator Training: A Systematic Literature Review

Muqaddas Abid, Mohammed Ismail Russtam Suhrab, Ahmad Faizal Ahmad Fuad, Muhammad Hammad, Shahrizan Jamaluddin, and Md Mahadi Hasan Imran

Abstract Training is vital across all industries, including maritime. Effective training is key to ensuring safe and efficient maritime operations. This study conducts a systematic literature review to explore factors that impact successful maritime simulator training programs. It identifies key factors, such as training material quality, trainer competence, realistic scenarios, and training frequency. Other significant factors include feedback mechanisms, innovative technology integration, and human factors in the curriculum. This study underscores the need for a comprehensive approach to maritime simulator training and advocates for advanced simulation technologies like virtual reality and artificial intelligence to enhance these programs. This literature review offers valuable insights and recommendations for improving maritime simulator training programs.

Keywords Factors · Maritime simulator · Training · Literature review

M. Abid · M. I. R. Suhrab (✉) · A. F. A. Fuad
Faculty of Maritime Studies, Universiti Malaysia Terengganu, 21030 Kuala Terengganu, Malaysia
e-mail: m.ismail@umt.edu.my

M. Abid
e-mail: muqaddasabid3@gmail.com

A. F. A. Fuad
e-mail: faizalfuad75@gmail.com

M. Hammad
Faculty of Business, Economics and Social Development, Universiti Malaysia Terengganu, 21030 Kuala Nerus, Terengganu, Malaysia
e-mail: fidahammad@gmail.com

S. Jamaluddin · M. M. H. Imran
Faculty of Ocean Engineering Technology, Universiti Malaysia Terengganu, 21030 Kuala Terengganu, Malaysia
e-mail: shahrizanj@umt.edu.my

M. M. H. Imran
e-mail: mmh.imran.official@gmail.com

A. Ismail et al. (eds.), *Advancements in Industry and Technology*,
SpringerBriefs in Applied Sciences and Technology,
https://doi.org/10.1007/978-3-031-97931-6_14

1 Introduction

The maritime profession is challenging due to sea conditions [1, 2]. Training is crucial for seafarers, with the International Maritime Organization (IMO) setting international standards through the Standards of Training, Certification, and Watchkeeping for Seafarers (STCW) since 1978 [3]. Maritime education and training (MET) provides seafarers with the necessary skills. MET includes theory-based education (classroom and textbooks) and practice-based education (simulators and sea experience). MET has a history of using simulator training for sailor competence [4].

Simulators aim to imitate vehicles or systems for training purposes [5]. They range from low to high fidelity, with low-fidelity simulators for basic training and high-fidelity simulators for advanced skills [4]. Institutions use four types of maritime simulators: Desktop-based, full mission bridge, virtual reality, and cloud-based [6]. Full mission bridge simulators are valuable for specialized training and research [7]. Virtual reality simulators provide immersive experiences [8], and cloud-based simulation allows online access [6]. The IMO foresees a shortage of well-trained sailors [9]. Simulator training in marine institutions imparts essential knowledge and expertise. This study focuses on factors contributing to successful simulator training in MET and their impact on training effectiveness.

2 Methodology

A systematic literature review, or SLR, is a research approach that seeks to answer research questions by reviewing the relevant body of literature surrounding a specific subject area. This paper aims to evaluate the breadth of scholarly research on the efficiency of simulator training. This study applies a literature review methodology to examine and synthesize complex, multidimensional data in maritime simulators for education and training and to explain findings in a manner that is accessible and understandable to a broad audience. This research adheres to the preferred reporting items for systematic reviews and meta-analyses (PRISMA) guidelines for reporting, which are the gold standard for such studies. The PRISMA statement intends to improve the transparency, dependability, and reproducibility of systematic reviews and meta-analyses by establishing reporting criteria and guidelines.

The current research methodology involves the following steps: (1) the use of keywords to locate relevant literature in scientific databases, (2) the application of inclusion and exclusion criteria to screen the literature, (3) the application of stage 2 screening to further scrutinize literature following the research question, and (4) the analysis of the finalized literature and conclusion.

2.1 Research Questions

This study focused on two core research questions which are as follows:

Research question 1: What are the factors contributing to the effectiveness of simulator training in maritime education and training?

Research question 2: What is the scope and extent of each factor toward the effectiveness of simulator training?

Reviewing the literature for question 1 will identify vital factors for improving maritime simulator training. Addressing question 2 will assist maritime researchers and instructors in enhancing training effectiveness. But first, the researcher wants to add the definition of effectiveness, which is, according to the Oxford Languages Dictionary, "*the degree to which something is successful in producing a desired result* [10]."

Search results were validated through test searches with different keyword permutations such as "simulator training" and "maritime simulators." An initial search was conducted on Scopus, Web of Science, and Springer using specific keywords (simulator OR simulation) AND (marine OR maritime) AND (training OR education). The study focused on articles and review articles published in English in the maritime domain from January 2018 to February 2023, resulting in 154 papers for initial screening.

2.2 Screening Phase

During the initial screening, we excluded 26 papers due to duplicates and the unavailability of full texts, leaving us with 128 papers. In the subsequent screening round, we carefully reviewed the abstracts and titles of these 128 papers to select those relevant to our research questions. This process resulted in 39 selected references for full-text examination. After a thorough reading, 12 articles were excluded for not addressing our research questions. As a result, we have 27 articles left for the final review. PRISMA framework and number of publications at each stage of the current study is shown in Fig. 1.

3 Results and Discussion

Maritime education research is needed to understand better how simulation training can help students prepare for the world's most safety-sensitive sector. This section will discuss the findings and conclusions of the current research. We will talk about each of the discovered factors one at a time, analyzing who worked on the project and the kind of research conducted. Figure 2 shows year-wise publications indicating that this area needs special attention for effective simulator training.

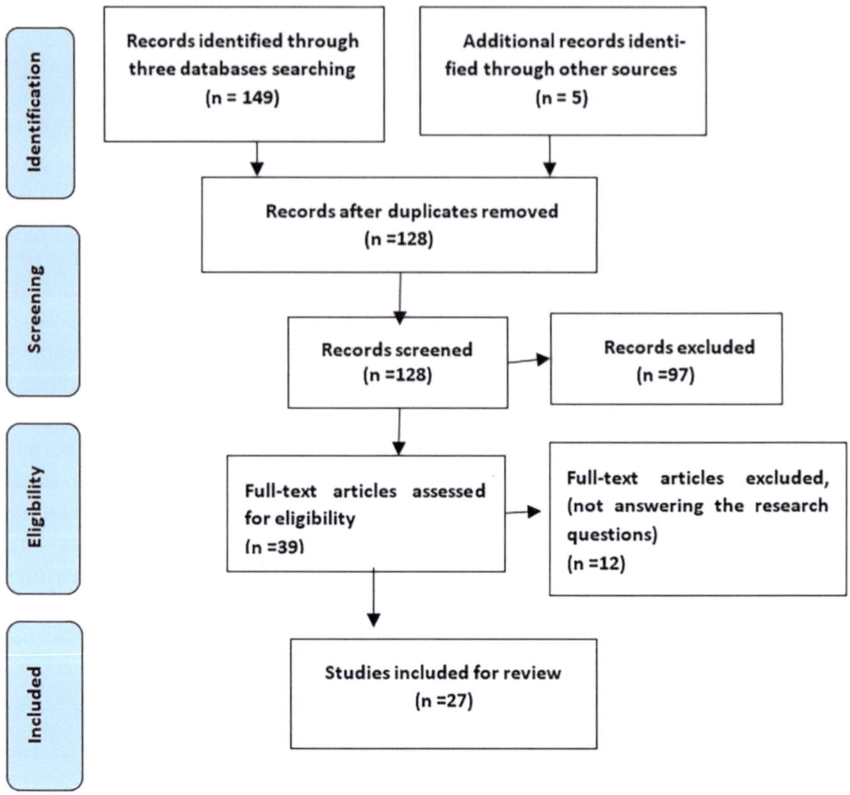

Fig. 1 PRISMA flowchart and the number of publications at the end of each step

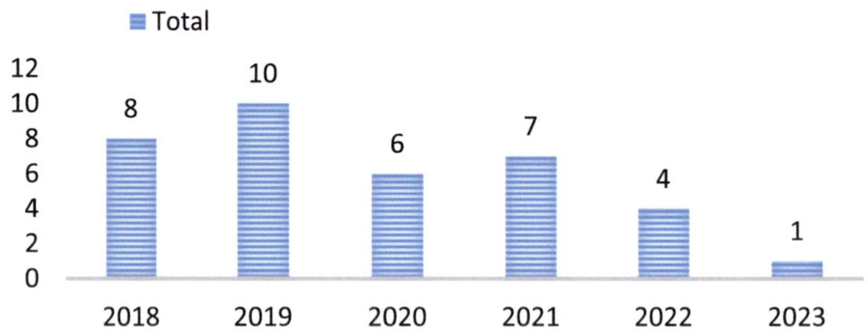

Fig. 2 Number of publications in the corresponding year

3.1 Simulation-Based Training (SBT) in MET

MARPOL, SOLAS, STCW, the ISM code, and the ISPS (International Ship and Port Security) code are significant marine industry agreements [11]. They outline provisions for process improvements and are endorsed by reputable maritime organizations and committees [12]. These agreements play a crucial role in enhancing the safety culture. Simulation, an interactive approach, enhances students' cognitive and psychomotor skills by providing a realistic learning environment [13]. Its use in the maritime industry, like in other training fields, is driven by technological, economic, and pedagogical needs [14]. When used effectively and supported by expert teachers [15], simulation helps students develop their skills and productivity in a risk-free setting. Simulation has significant pedagogical value in maritime education, aiding in the training of competent sailors. It also enables the evaluation of routes, weather, and traffic, and the management of potentially hazardous situations without real-world risks [16]. Effective training programs and design, along with the instructor's role, contribute to the success of simulator-based training.

3.2 *The Organization of Simulator-Based Training*

Simulator-based training typically consists of three stages: briefing, scenario, and debriefing [17]. The briefing phase introduces the day's tasks, focusing on practical knowledge, resource sharing, and learning objectives [18, 19]. Instructors use classroom tools like documents and presentations to provide clear instructions. Following the briefing, a scenario is conducted in the simulator. The maritime instructor assigns student teams to ship's bridge duties, observing their activities closely to assess alignment with learning goals [17]. The final stage is a debriefing session, where participants reflect on the experience and use feedback technologies for insights and improvement, common in aviation and marine training. This phase clarifies specific details for safe navigation [17].

3.3 *Role of Assessment and Its Methods in SBT*

The significance of marine simulation training mandates institutes to adopt a more effective approach to teaching and assessing students. Previous research has highlighted key aspects related to using simulation in marine education and training [4, 6, 11, 20]. As recommended by [21], student evaluation should cover three domains: cognitive, psychomotor, and affective. Successful training programs rely on robust performance assessment, necessitating reliable and valid measurements, as emphasized by [17, 22] underscore the importance of assessment and feedback in simulator-based training, revealing issues related to inadequate simulator use and ineffective evaluations. According to [23], a thorough assessment method for trainee evaluation, emphasizing its crucial role in achieving effective training outcomes. Additionally, [22] introduces an assessment tool for maritime simulation training, focusing on the assessment criteria's significance in developing skilled seafarers. In another study,

[24] discusses the need for objective and transparent assessment and emphasizes the importance of valid and reliable competency tests at the training's conclusion.

3.4 Role of Instructor in Simulation-Based Training

Simulator training's simplicity and adaptability are driving demands [25]. Effective maritime training hinges on competent instructors, who must design scenarios, assign tasks, and ensure learning transfer in emergencies and response operations [26]. Elevated instructor certification levels are essential for maximizing simulator potential [23], with professional supervision highlighted in [17]. Instructors play a pivotal role, acting as facilitators during simulator exercises and debriefings [18], fostering learning, confidence, teamwork, and safety awareness for trainees' effective and safe job performance.

3.5 Training Design

The maritime industry increasingly relies on simulators for training, offering a safe and controlled environment for complex tasks, error reduction, and performance improvement in high-risk situations. Training design is pivotal, involving the creation of new programs, materials, and lessons, and identifying gaps to enhance performance. It is crucial to gradually increase scenario complexity [27], as sudden exposure can hinder performance. Starting with simpler tasks and gradually elevating complexity aligns with findings by [28] regarding training design, particularly for docking operations. According to [25], a well-planned development of exercise objectives, target-focused facilitation, and formal exercise debriefing and evaluation are considered essential components of successful training.

3.6 Training Time

Simulator training is vital in the maritime industry, with the number of hours spent in a simulator being a critical factor for its effectiveness. Research shows a direct link between training hours and real-life performance, as trainees need time to adapt and immerse themselves in the simulation [6]. Allowing participants to acclimatize to the simulated environment aids in assessing workload, stress, and situational control. However, students often receive insufficient simulator training during their education [28], and the optimal duration for effective simulator use is still being explored. Enhancing the learning of COLREGs through navigation simulators is crucial [29], emphasizing the need to invest an adequate number of training hours to

Table 1 Summary of search results

Database	Total entries found	Article and review article in English	Source/Publication title
Web of science	101,169	64,380	60
Springer	1934	267	30
Scopus	493	286	59
Total			149

ensure individuals acquire the essential skills and knowledge for proficient real-life performance.

3.7 *Simulator Technology/Simulator Fidelity*

In the maritime industry, simulator fidelity, which reflects the accuracy and realism of a simulation, is increasingly crucial. Replicating real-world conditions precisely is essential for effective and reliable simulations. Studies emphasize the importance of matching the fidelity level to the specific training activity, with a focus on functional fidelity [27], and how realistically the simulator responds to trainees, rather than physical fidelity. Moreover, the level of fidelity should match the specific activity for which the subject is being trained. Simulators play a valuable instructional role, with authenticity, the resemblance between the simulation and real work settings, aiding the transfer of skills and knowledge to the workplace [20]. However, achieving high accuracy in the simulator's physical properties does not guarantee effective learning [30], as a low level of realism can limit the training experience [31, 32]. This limitation may cause participants to become overly focused on manipulating the simulator rather than gaining practical real-world experience. While some research suggests that high-fidelity simulators yield better learning outcomes [30], the relationship between simulator realism and learning is still in its early stages (Table 1).

4 Conclusion

Maritime simulator training is a crucial component of maritime education, offering a safe and cost-effective approach to prepare seafarers for various real-life scenarios at sea. However, a recent systematic literature review highlighted the scarcity of research examining the factors contributing to effective simulator training. Instructors play a pivotal role in success, while realistic scenarios, feedback, and tailored programs are key. Customized training durations may be needed for proficiency and safety. In conclusion, maritime simulator training is efficient but needs more research for improvement.

References

1. A. Androjna, T. Brcko, I. Pavic, H. Greidanus, Assessing cyber challenges of maritime navigation. J. Mar. Sci. Eng. **8**, 1–21 (2020)
2. J. Banas, T. Mehling, T. Paul, A. Cords, Manned-unmanned teaming challenges in the maritime environment, in *76th Annual Forum of the Vertical Flight Society*, vol. 4 (2020)
3. M. H. Dewan, R. Godina, M. R. K. Chowdhury, C. W. M. Noor, W. M. N. Wan Nik, M. Man, Immersive and non-immersive simulators for the education and training in maritime domain—a review. J. Mar. Sci. Eng. **11**, 147 (2023)
4. S.K. Renganayagalu, S.C. Mallam, M. Hernes, Maritime education and training in the COVID-19 era and beyond. TransNav **16**, 59–69 (2022)
5. Oxford Dictionary. https://www.oxfordlearnersdictionaries.com/definition/english/simulator (2023). Accessed 8 May 2023
6. T.E. Kim, A. Sharma, M. Bustgaard, W.C. Gyldensten, O.K. Nymoen, H.M. Tusher, S. Nazir, The continuum of simulator-based maritime training and education. WMU J. Marit. Aff. **20**, 135–150 (2021)
7. V. Senčila, R. Zažeckis, A. Jankauskas, R. Eitutis, The use of a full mission bridge simulator ensuring navigational safety during the Klaipeda seaport development. TransNav **14**, 417–424 (2020)
8. J. Radianti, T. A. Majchrzak, J. Fromm, I. Wohlgenannt, A systematic review of immersive virtual reality applications for higher education: design elements, lessons learned, and research agenda. Comput. Edu. **147** (2020).
9. D. Ivanišević, A. Gundić, D. Mohović, Difficulties in understanding the COLREGs among the students from different systems of education for seafarers. TransNav. **13**, 869–873 (2019)
10. Oxford Languages. https://www.google.com/search?q=definition+of+simulator (2023). Accessed 8 May 2023
11. M. Baldauf, D. Dalaklis, A. Kataria, Team training in safety and security via simulation: a practical dimension of maritime education and training, in *The Maritime Commons: Digital Repository of the World Maritime University* (IATED Academy, 2016)
12. G. Kara, O.H. Arıcan, O. Okşaş, Analysis of the effect of electronic chart display and information system simulation technologies in maritime education. Mar. Technol. Soc. J. **54**, 43–57 (2020)
13. M.L. Rhodes, C. Curran, Use of the human patient simulator to teach clinical judgment skills in a baccalaureate nursing program. Comput. Inform. Nurs. **23**, 256–262 (2005)
14. E. Çakir, Ö. Arslan, Turkish telemedical assistance service: last four years of activity. Int. Marit. Health **69**, 184–191 (2018)
15. A.A. Malik, N. Zafar, Applications of simulation technology—pitfalls and challenges. TransNav, Int. J. Mar. Nav. Saf. Sea Transp. **9**, 391–396 (2015)
16. K. Ivancic, B. Hesketh, Learning from errors in a driving simulation: effects on driving skill and self-confidence. Ergonomics **43**, 1966–1984 (2000)
17. C. Sellberg, From briefing, through scenario, to debriefing: the maritime instructor's work during simulator-based training. Cogn. Technol. Work **20**, 49–62 (2018)
18. R. M. Fanning, D. M. Gaba, The role of debriefing in simulation-based learning. Simul. Healthc. **2**, 115–125 (2007). https://doi.org/10.1097/SIH.0b013e3180315539
19. M.P. Wickers, Establishing the climate for a successful debriefing. Clin. Simul. Nurs. **6**, 83–86 (2010)
20. S. C. Mallam, S. Nazir, S. K. Renganayagalu, Rethinking maritime education, training, and operations in the digital era: applications for emerging immersive technologies. J. Mar. Sci. Eng. **7**, 1–9 (2019). https://doi.org/10.3390/JMSE7120428
21. S. Jeanne Horst, C. O. Prendergast, The assessment skills framework: a taxonomy of assessment knowledge, skills and attitudes. J. Res. Pract. Assess. **15**(1) (2014)
22. J. Ernstsen, S. Nazir, Performance assessment in full-scale simulators—a case of maritime pilotage operations. Saf Sci **129** (2020)

23. C. Kandemir, O. Soner, M. Celik, Proposing a practical training assessment technique to adopt simulators into marine engineering education. WMU J. Marit. Aff. **17**, 1–15 (2018)
24. C. Sellberg, Pedagogical dilemmas in dynamic assessment situations: perspectives on video data from simulator-based competence tests. WMU J. Marit. Aff. **19**, 493–508 (2020)
25. J. Halonen, A. Lanki, Efficiency of maritime simulator training in oil spill response competence development. TransNav **13**, 199–204 (2019)
26. J.J. Victoriano, R.T. Villaruz, R.A. Alimen, Teaching situation and instructors' key-competencies at the college of maritime education (CME): evaluation of the marine engineering students. TransNav **16**, 241–250 (2022)
27. K. Hjelmervik, S. Nazir, A. Myhrvold, Simulator training for maritime complex tasks: an experimental study. WMU J. Marit. Aff. **17**, 17–30 (2018)
28. S. Nazir, S. Jungefeldt, A. Sharma, Maritime simulator training across Europe: a comparative study. WMU J. Marit. Aff. **18**, 197–224 (2019)
29. C. Sellberg, O. Lindmark, H. Rystedt, Learning to navigate: the centrality of instructions and assessments for developing students' professional competencies in simulator-based training. WMU J. Marit. Aff. **17**, 249–265 (2018)
30. S.K. Renganayagalu, S.C. Mallam, S. Nazir, J. Ernstsen, P. Haavardtun, Impact of simulation fidelity on student self-efficacy and perceived skill development in maritime training. TransNav **13**, 663–669 (2019)
31. M. Baric, J. Ēulin, T. Bielic, Problems that occur in a team: learning from maritime accidents via simulation training. TransNav, Int. J. Mar. Navig. Saf. Sea Transp. **12**, 709–713 (2019)
32. J.I. Alcaide, Modelling the relationship between performance and ship-handling simulator. J. Marit. Res. **17**, 68–73 (2020)

Assessment and Mitigation of Seafarer's Fatigue Among Singapore Bunker Tanker Crews

Md Ranakul Islam, Mohammed Ismail Russtam Suhrab, Ahmad Faizal Ahmad Fuad, and Rasyidah Mohd Sani

Abstract In the bustling waters of Singapore, seafarers on bunker tankers endure relentless fatigue, exposing challenges in the shipping industry. This research delves into fatigue, exploring its facets from ship design to crew well-being. Sleep deprivation resonates like a siren, shaping their fatigue amidst crashing waves and quiet nights. Through a literature review and empowered by Bayesian networks and Python software, the study navigates data collection, seeking insights. The battle proves tough, with physical and psychological stress and shore-based management factors as adversaries. Nevertheless, hope persists; means to protect these courageous souls are within reach. Armed with newfound knowledge, the research sails toward a brighter horizon, easing their fatigue and offering solace in the tempestuous seas of Singapore's shipping industry.

Keywords Seafarers · Bunker tankers · Fatigue · Singapore · Shipping industry

1 Introduction

Seafaring is a challenging profession that plays a crucial role in the global economy, facilitates the passage of products and energy resources across the world's oceans [1]. Among the numerous maritime operations, Singapore's bunker tanker crews hold a unique and essential position, responsible for the proper refueling of ships which

M. R. Islam · M. I. R. Suhrab (✉) · A. F. A. Fuad · R. M. Sani
University Malaysia Terengganu, Kuala Terengganu, Terengganu, Malaysia
e-mail: m.ismail@umt.edu.my

M. R. Islam
e-mail: ranakulmariner@gmail.com

A. F. A. Fuad
e-mail: afaf@umt.edu.my

R. M. Sani
e-mail: s55177@umt.edu.my

A. Ismail et al. (eds.), *Advancements in Industry and Technology*,
SpringerBriefs in Applied Sciences and Technology,
https://doi.org/10.1007/978-3-031-97931-6_15

ensures the continuous flow of goods and commodities [2, 3]. These enthusiastic seafarers operate in an environment where the unrelenting demands of the industry collide with the harsh obstacles of the open sea. Nevertheless, considering these duties and hardships, seafarer exhaustion poses substantial threats to the safety, well-being, and environmental integrity of these essential operations.

Fatigue among seafarers is a prevalent issue that impacts not just individuals operating ceaselessly on vessels, but also maritime safety and sustainability on a greater level [4]. Seafarers become fatigued and make serious mistakes in decision-making procedures which lead to incidents, because of which the whole world needs to pay the price [5, 6]. Such as, the Exxon Valdez oil spill occurred on March 24, 1989, in Prince William Sound, Alaska, causing nearly 11 million US gallons of crude oil to spill over several days. The cleanup and related civil and criminal charges incurred a staggering cost of approximately $3 billion for Exxon [7–11].

This study investigates seafarer fatigue among those working on Singapore's bunker tanker vessels with a dual focus: firstly, to enhance safety and prevent accidents by identifying and understanding the factors contributing to fatigue, thus safeguarding crew well-being and the environment in one of the world's busiest maritime hubs, and secondly, observe the fatigue level and finally proper recommendation to minimize the fatigue level.

2 Literature Review

In the maritime sector, there is a common tendency to attribute safety issues on commercial ships to "human error" [12]. However, investigations into the root causes of maritime accidents reveal that this assertion lacks complete accuracy [13]. Over the past few decades, the maritime industry has witnessed a significant increase in globalization and seaborne trade, driven by technological advancements and automation, which have substantially reduced the number of crew members on commercial vessels [14]. Simultaneously, the persistence of shipping accidents has led to the introduction of more stringent safety regulations by the International Maritime Organisation (IMO). While these regulations are essential, their implementation has imposed additional responsibilities on the already understaffed crews, resulting in heightened fatigue levels [15].

Human fatigue is a multifaceted issue with significant implications for physical and mental health, often associated with long-term consequences. Although there is no universally accepted definition of fatigue in the literature, it is frequently described as a subjective state where individuals lack confidence in their ability to efficiently perform tasks, often stemming from inadequate physical rest and diminished alertness. Extensive research has consistently linked fatigue to accidents, underscoring its critical role in safety [16, 17].

The analysis of maritime accidents consistently highlights the significant role of fatigue in contributing to these incidents [18–20]. The dynamic nature of maritime operations necessitates seafarers to work at varying times of the day, leading to sleep deprivation and circadian rhythm disruptions [15, 21]. Past research indicates that the performance of maritime workers can vary throughout the day and decline with extended work hours [22]. However, the subjective nature of fatigue makes it challenging to quantify and objectively demonstrate its influence on accidents, rendering an examination of individual causal factors insufficient for understanding its contribution. Therefore, it becomes crucial to delve into what motivates seafarers to work under the influence of fatigue, prompting the design of a qualitative study to analyze firsthand accounts of seafarers involved in such accidents.

3 Methodology

To conduct this study, a questionnaire-based survey method was used to collect data. Firstly, we developed survey questionaries' using Google Forms and subsequently distributed them among seafarers who were working in Singapore bulk carriers using the stratified complex random sampling process. Data analysis was carried out using Bayesian methods once we received their responses. Three research questions were formed: RQ1: What are the factors contributing to fatigue in bunker crews? RQ 2: What is the level of fatigue of bunker crews? and RQ3: What are the measures to minimize the fatigue level for bunker crews?, Furthermore, various factors were identified that could impact their fatigue levels, including, ship design quality (SDQ), health quality (HQ), sleep quality (SQ), ashore management (AM), physical stress level (PS), psychological stress level (PXL), and fatigue level (FL) of seafarers.

4 Result and Discussion

Correlation Matrix Analysis

Figure 1 shows the correlation matrix between ship design quality and the fatigue level. The correlation between ship design quality (SDQ) and fatigue level (FL) showed no significant impact on fatigue, with a p-value of 0.9305. However, sleep quality (SQ) had a significant effect on fatigue, as indicated by a p-value of 0.016 and a t-value of 2.44, suggesting a difference in fatigue levels between those with good sleep and rest quality and those without. Similarly, physical stress (PS) and psychological stress (PXL) were found to have a significant correlation with fatigue level, both with a p-value of 0.0029 and t-values of 3.05, indicating higher fatigue levels for those experiencing high levels of stress.

Critical Factors Analysis

Ship design quality (SDQ) received high ratings, with 87 out of 95 values indicating good perceptions. Health quality (HQ) and sleep quality (SQ) were generally positive, as most crew members reported good health and normal sleep quality. Ashore

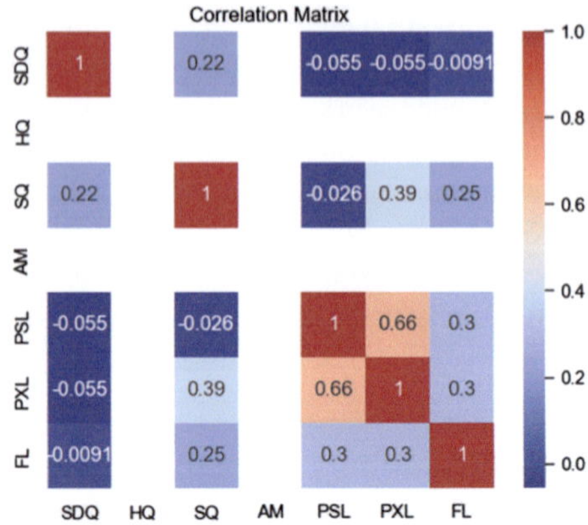

Fig. 1 Correlation matrix between ship design quality and the fatigue level

management (AM) received no negative feedback, suggesting it did not contribute to fatigue. Physical stress level (PSL) and psychological stress level (PSXL) were within normal ranges for the majority of the crew (92 out of 95). However, fatigue level (FL) indicated some level of fatigue in most crew members, with 70 out of 95 reporting it, while 25 reported no fatigue, signifying that fatigue was a prevalent concern among crew members. Figures 2, 3, 4, and 5 show the frequency plots of the various factors leading to fatigue levels among seafarers.

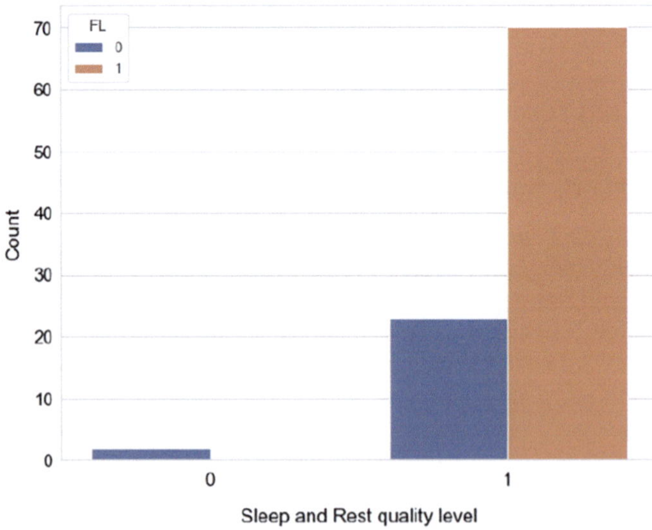

Fig. 2 Frequency plots of sleep and rest quality level

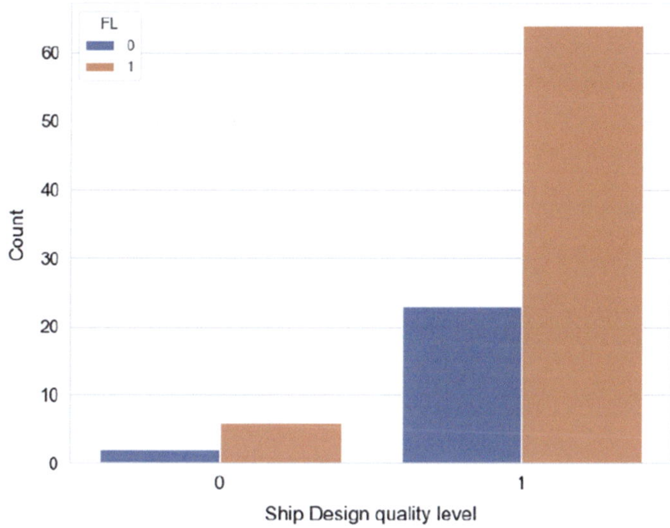

Fig. 3 Frequency plots of ship design quality level

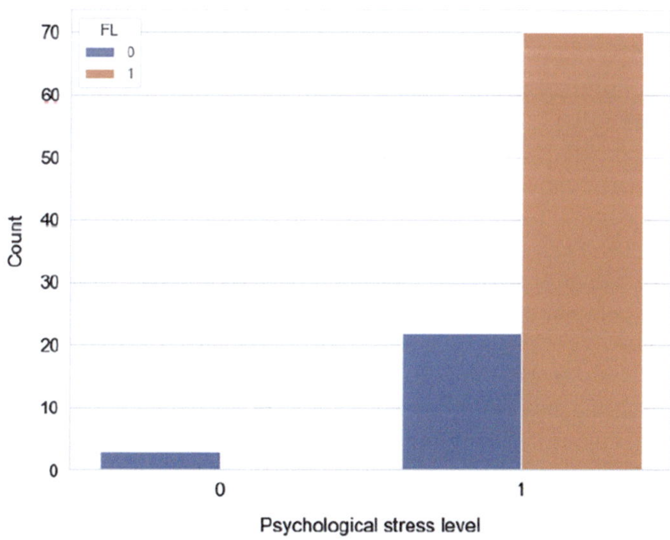

Fig. 4 Frequency plots of psychological stress level

Bayesian Estimator Analysis

The output Fig. 6 shows the values of the joint probability distribution for six different variables: AM, PSL, SDQ, SQ, HQ, and FL. The values in the table are the probabilities of all possible combinations of values for the six variables. For example, the

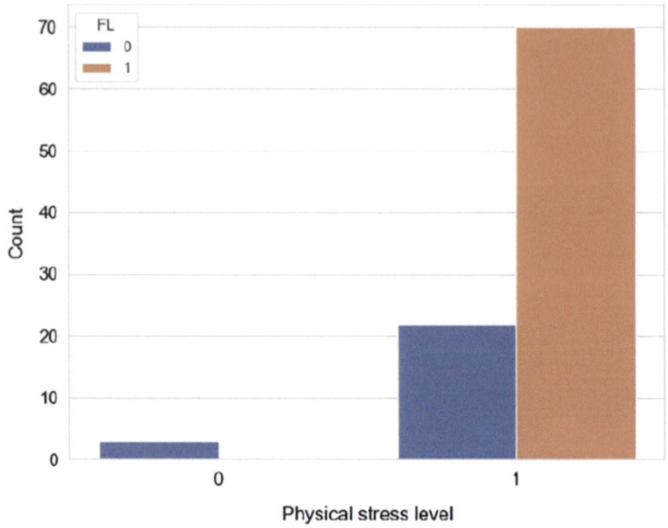

Fig. 5 Frequency plots of physical stress level

first row of the table corresponds to the case where all six variables have the value 0. The probability of this case is 0.0007, which means that this case is expected to occur with a frequency of 0.07% in the data. Similarly, the third row corresponds to the case where only the SL variable has the value 1, and all other variables have the value 0. The probability of this case is also 0.038, which means that this case is also expected to occur with a frequency of 3.80% in the data.

The other rows in the table correspond to all possible combinations of values for the six variables. For each row, the probability of that combination is listed in the last column of the table. The joint probability distribution is useful for analyzing the relationships between different variables. For example, it can be used to calculate the conditional probabilities of certain events given the occurrence of other events. The last row, for example, shows that fatigue level 1 occurs along with sleep quality of 1 and physical stress quality of 1, leading to a probability of 62.5% of the case occurring. This shows a strong correlation between bad sleep and high physical stress leading to high fatigue levels.

Maximum Likelihood Estimator

Pythod prediction demonstrates that the most impactful features for fatigue levels are physical stress and sleep quality. 10.5% probability of ship design quality to be in state 0 and 89.5% to be in state 1. The model predicts that there is a 4.5% probability of sleep quality being in state 0 and 95.5% of being in state 1. a 5.5% probability of physical and psychological stress being in state 0 and 94.5% being in state 1. Figure 6 shows the most impactful features for fatigue levels, namely physical stress and sleep quality.

AM	PSL	SDQ	SQ	HQ	FL	phi(AM,PSL,SDQ,SQ,HQ,FL)
AM(0)	PSL(0)	SDQ(0)	SQ(0)	HQ(0)	FL(0)	0.0007
AM(0)	PSL(0)	SDQ(0)	SQ(0)	HQ(0)	FL(1)	0.0007
AM(0)	PSL(0)	SDQ(0)	SQ(1)	HQ(0)	FL(0)	0.0038
AM(0)	PSL(0)	SDQ(0)	SQ(1)	HQ(0)	FL(1)	0.0038
AM(0)	PSL(0)	SDQ(1)	SQ(0)	HQ(0)	FL(0)	0.0056
AM(0)	PSL(0)	SDQ(1)	SQ(0)	HQ(0)	FL(1)	0.0056
AM(0)	PSL(0)	SDQ(1)	SQ(1)	HQ(0)	FL(0)	0.0324
AM(0)	PSL(0)	SDQ(1)	SQ(1)	HQ(0)	FL(1)	0.0031
AM(0)	PSL(1)	SDQ(0)	SQ(0)	HQ(0)	FL(0)	0.0030
AM(0)	PSL(1)	SDQ(0)	SQ(0)	HQ(0)	FL(1)	0.0004
AM(0)	PSL(1)	SDQ(0)	SQ(1)	HQ(0)	FL(0)	0.0154
AM(0)	PSL(1)	SDQ(0)	SQ(1)	HQ(0)	FL(1)	0.0773
AM(0)	PSL(1)	SDQ(1)	SQ(0)	HQ(0)	FL(0)	0.0148
AM(0)	PSL(1)	SDQ(1)	SQ(0)	HQ(0)	FL(1)	0.0143
AM(0)	PSL(1)	SDQ(1)	SQ(1)	HQ(0)	FL(0)	0.1936
AM(0)	PSL(1)	SDQ(1)	SQ(1)	HQ(0)	FL(1)	0.6256

Fig. 6 Values of the joint probability distribution for AM, PSL, SDQ, SQ, HQ, and FQ

5 Recommendations

According to the analysis, it has demonstrated that fatigue levels among crew members are influenced by physical and psychological stress. To manage fatigue, organizations can implement interventions such as evenly distributed tasks, proper training, and teamwork. Personal support can be provided through mentors, physiotherapists, or counselors, while family issues can be addressed professionally. Realistic performance expectations should be set, considering long-term organizational needs. Leaves should be rechargeable and recognized as a crucial aspect of compensation. A supportive work environment with strong relationships between supervisors and crew members can promote stability and well-being. By implementing these interventions, organizations can effectively manage fatigue levels, improve overall well-being, and create a safe and productive work environment.

6 Conclusion

The seafaring industry is crucial for global trade and should prioritize the physical, mental, and emotional well-being of its workers. Emotional health at the workplace is essential for preventing fatigue, as it plays a significant role in amplifying fatigue levels. The post-Covid era has highlighted the importance of rest and recovery, as the pandemic has caused livelihood disruptions, separation from family, income loss, and job loss, contributing to increased fatigue levels among seafarers. Research on fatigue, particularly focusing on tanker vessel seafarers, is necessary to prevent accidents and promote safety.

References

1. R. C. Saha, The role of Chittagong port authority to develop other national ports in Bangladesh to provide maritime logistics support in South Asia. Mar. Struct. **5** (2023). https://doi.org/10.36956/sms.v5i1.824
2. N. Gray, S. McDonagh, R. O'Shea, B. Smyth, J. D. Murphy, Decarbonising ships, planes and trucks: an analysis of suitable low-carbon fuels for the maritime, aviation and haulage sectors. Adv. Appl. Energy **1** (2021)
3. A. Rajapakse, G. R. Emad, Fatigue, an unsolved puzzle that continues contributing to accidents at sea. Mar Policy. **155** (2023). https://doi.org/10.1016/j.marpol.2023.105745
4. N. Turgo, Manning the waves: masculinity shift amongst Filipino seafarers in the age of precarity. Gend. Work Organ. **28** (2021). https://doi.org/10.1111/gwao.12710
5. Y. Jiao, M. A. Dulebenets, Y. Y. Lau, Cruise ship safety management in Asian regions: trends and future outlook. Sustain. (Switz.). **12** (2020). https://doi.org/10.3390/su12145567
6. S. Fan, E. Blanco-Davis, S. Fairclough, J. Zhang, X. Yan, J. Wang, Z. Yang, Incorporation of seafarer psychological factors into maritime safety assessment. Ocean Coast Manage. **237** (2023). https://doi.org/10.1016/j.ocecoaman.2023.106515
7. Z. Asif, Z. Chen, C. An, J. Dong, Environmental impacts and challenges associated with oil spills on shorelines, J. Mar. Sci. (2022)
8. R. Sharma, N. S. Singh, N. Dhingra, T. Parween, Bioremediation of oil-spills from shoreline environment, in *Modern Age Waste Water Problems* (2020)
9. B. Shyam, R. Phukan, C. Das, Bioremediation of oil spill cleanup: case studies of two major oil spills, in *Sustainable Environment* (2023)
10. A. Juhasz, *Investigation: Two Years After the BP Spill, A Hidden Health Crisis Festers* (2012)
11. S. Haycox, Crude oil pollution III. Exxon Valdez contamination: ecological recovery, a case study, in *A Handbook of Environmental Toxicology: Human Disorders and Ecotoxicology* (2020)
12. M. Celik, S. M. Lavasani, J. Wang, A risk-based modelling approach to enhance shipping accident investigation. Saf Sci. **48** (2010). https://doi.org/10.1016/j.ssci.2009.04.007
13. F. Van De Merwe, N. Kähler, P. Securius, Crew-centred design of ships—the CyClaDes project, in *Transportation Research Procedia* (2016)
14. M. R. Grech, Fatigue risk management: a maritime framework. Int J Environ Res Public Health. **13** (2016). https://doi.org/10.3390/ijerph13020175
15. M. Lützhöft, B. Thorslund, A. Kircher, M. Gillberg, *Fatigue at Sea: A Field Study in Swedish Shipping* (VTI rapport, 2007)
16. J. Dorrian, S. D. Baulk, D. Dawson, Work hours, workload, sleep and fatigue in Australian rail industry employees. Appl Ergon. **42** (2011). https://doi.org/10.1016/j.apergo.2010.06.009

17. O. C. Jensen, J. F. L. Sørensen, M. L. Canals, Y. P. Hu, N. Nikolic, M. Thomas, Incidence of self-reported occupational injuries in seafaring—an international study. Occup Med (Chic Ill). **54** (2004). https://doi.org/10.1093/occmed/kqh090
18. M. J. Akhtar, I. B. Utne, Human fatigue's effect on the risk of maritime groundings—a Bayesian Network modeling approach. Saf Sci. **62** (2014). https://doi.org/10.1016/j.ssci.2013.10.002
19. N. A. Costa, E. Holder, S. N. MacKinnon, Implementing human centred design in the context of a graphical user interface redesign for ship manoeuvring. Int. J. Hum. Comput. **100** (2017). https://doi.org/10.1016/j.ijhcs.2016.12.006
20. B. Strauch, Investigating fatigue in marine accident investigations. Procedia Manuf. **3** (2015). https://doi.org/10.1016/j.promfg.2015.07.859
21. A. Rajapakse, G. R. Emad, M. Lützhöft, M. Grech, A study on time constraints and task deviations at sea leading to accidents–a cultural-historical perspective. Marit. Policy Manage. **46** (2019). https://doi.org/10.1080/03088839.2019.1574407
22. P. Boudreau, S. Lafrance, D. B. Boivin, Alertness and psychomotor performance levels of marine pilots on an irregular work roster. Chronobiol. Int. **35** (2018). https://doi.org/10.1080/07420528.2018.1466796

AI and Access to Justice: Digital Gap Debunked?

Ahmad Hafiz Zubir and Ida Madieha Azmi

Abstract Artificial intelligence (AI) is transforming the legal field, with significant implications for access to justice. This research paper delves into the role of AI in legal practice, particularly its effects on due process, legal representation, liability, and the broader accessibility of legal services. Utilizing a methodological approach that combines a survey of legal practitioners with a review of recent studies, this paper evaluates how AI technologies might either bridge or further widen the existing digital divide in the legal sector. The results indicate that while AI significantly enhances efficiency in legal research and document drafting, it often falls short in performing complex legal analyses, which could adversely affect the due process and the quality of legal representation. The study underscores the necessity of developing a balanced regulatory framework to ensure AI is used ethically and effectively in legal practice, ensuring that it supports rather than undermines justice.

Keywords Artificial intelligence · Legal practice · Digital gap · Legal transformation · Legal challenges

1 Introduction

The integration of artificial intelligence (AI) into the legal sector is reshaping how legal services are delivered, enhancing efficiency and broadening access. AI technologies, capable of tasks traditionally requiring human intelligence such as data analysis and routine task automation, are becoming increasingly crucial as legal

A. H. Zubir (✉) · I. M. Azmi
International Islamic University of Malaysia, Gombak, Selangor, Malaysia
e-mail: hafiz@hzac.com.my

I. M. Azmi
e-mail: imadieha@iium.edu.my

A. H. Zubir
Hafiz Zubir & Co, Bandar Baru Bangi, Selangor, Malaysia

A. Ismail et al. (eds.), *Advancements in Industry and Technology*,
SpringerBriefs in Applied Sciences and Technology,
https://doi.org/10.1007/978-3-031-97931-6_16

systems worldwide struggle with growing caseloads and data management demands [1–3].

Historically slow to adopt new technologies, the legal profession has seen a shift as AI begins to streamline operations from case prediction to document automation, significantly impacting workflow and potentially enhancing service delivery [4–8]. However, the adoption of AI raises profound ethical questions about the due process, equity, and transparency in legal outcomes [6, 9–12].

AI tools, while reducing human workload and improving procedural efficiency, also pose risks of perpetuating existing biases or introducing new injustices if not judiciously managed. This is particularly concerning in terms of access to justice, where AI's potential to either bridge or widen the digital (gap) divide is still under intense scrutiny [5, 13, 14].

This paper will delve into the dual-edged nature of AI in legal practice, exploring both its transformative impacts and the significant challenges it poses. Drawing on a comprehensive review of recent studies and reports, the paper will discuss AI's role in legal workflows, its implications for access to justice, and the necessary considerations for its ethical deployment.

As AI becomes a staple in legal practice, understanding its implications on the justice system is crucial. This research contributes to the ongoing dialogue on AI in the legal field, advocating for robust regulatory frameworks that ensure AI supports rather than undermines justice [15–19].

2 Methodology

This study employed a mixed-methods approach to comprehensively assess the impact of artificial intelligence on legal practices and its broader implications for access to justice. The primary objective was to explore both the beneficial and challenging aspects of AI integration in legal settings.

2.1 Participants

The survey involved participants from various sectors of the legal industry, including practitioners and in-house lawyers, with a focus on those who have interacted with or implemented AI tools in their practice.

Participants varied in their level of experience, from novices in the early stages of their careers to seasoned professionals with over twenty years in the field. Participants were selected based on their direct involvement with AI technologies, ensuring a rich and relevant data pool.

2.2 Data Collection Methods

Data was collected primarily through a structured online survey distributed to legal professionals. The survey included a range of questions designed to gauge the participants' experiences with AI in legal work, their perceptions of its effectiveness, and the ethical concerns it raises.

The questionnaire was designed to assess both qualitative and quantitative aspects of AI use in legal practices. Questions were structured in open-ended questions allowing for detailed comments and insights. Additional data was gathered through follow-up interviews with a select group of respondents to deepen the insights obtained from the surveys.

2.3 Data Analysis

Quantitative data from the survey was analyzed using statistical methods to identify patterns and correlations in the responses. Qualitative data from open-ended survey questions and interviews were analyzed thematically to extract common themes related to the benefits, challenges, and ethical considerations of using AI in legal practice.

2.4 Ethical Considerations

The study was conducted in accordance with ethical research standards. All participants were informed of the study's purpose and their rights, with informed consent obtained before participation.

3 Results and Discussion

The survey conducted among legal practitioners revealed a diverse range of opinions and practices regarding the use of AI in legal work. Key quantitative findings indicate that a significant portion of respondents have engaged with AI tools in their practices, particularly in areas such as legal research, document drafting, and contract review.

A majority of respondents (57%) indicated that they use AI tools like ChatGPT and other legal technology software to enhance their workflow efficiency. Common applications mentioned include automated document analysis and predictive analytics for litigation outcomes.

While many respondents acknowledged the efficiency gains from AI, there was a notable concern about AI's depth of analysis. For example, one respondent noted, "AI

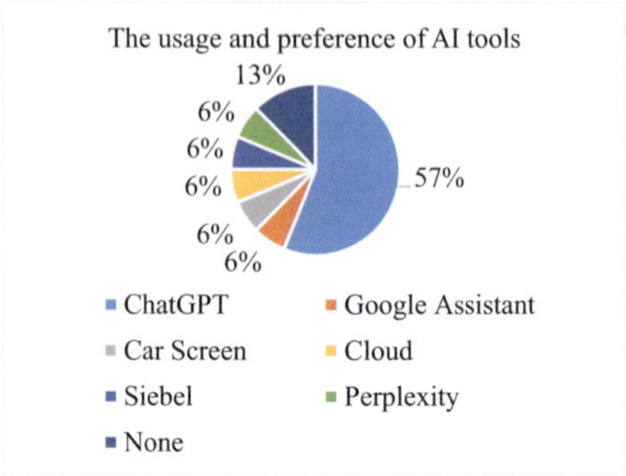

Fig. 1 The usage and preference of AI tools

output lacks depth of analysis which is critical to the legal profession," highlighting skepticism about AI's ability to fully replicate nuanced legal reasoning.

Ethical dilemmas were frequently mentioned, with concerns about data privacy, the potential for bias in AI algorithms, and the lack of control over automated processes. Respondents were particularly cautious about AI's role in client representation and court appearances, emphasizing the need for human oversight.

The survey indicated a perceived gap in current legislation related to AI in legal practice. Respondents suggested that Malaysian law lacks specific provisions for regulating AI, with one stating that MOSTI is planning to use the EU AI Act as a baseline. Below are six highlighted findings derived from the survey that has been converted into charts to help depict the nuances of answers by respondents in regard to AI use.

One of the earlier questions asked to the respondents was about the type of AI application the respondents were currently using. This question was posed to find out which of the many publicly available AI tools is the most used and the most helpful to the people in the legal profession. As observed from Fig. 1, ChatGPT has become the most preferred tool to assist the respondents in their legal works.

Figure 2 shows the usual function associated or used by the respondents in regard to the services offered by the AI tools. Based on the chart above, large percentage of respondents have been using the AI service in researching and drafting for the sake of saving time. Smaller percentage of them use the AI service to analyze a certain set of information and to come up with a new one. The least service utilized is in getting a particular guideline or perspective over an issue.

In general, Fig. 3 shows that most respondents have a neutral stance over the usage of AI in the legal profession. Despite some of them, especially those who have been in the legal field for more than 10 years, disapprove such a usage, they do not believe that using AI is wrong. Ethically, most respondents accepted its use in the practice.

Following the review on the respondents' most used AI tools and in what matter do they use the tools for, Fig. 4 shows a high percentage of respondents having a positive outlook on AI role in the access to justice. Only 20% of the respondent believed that AI will injure the access due to the lack of human touch or compassion in its generated analysis.

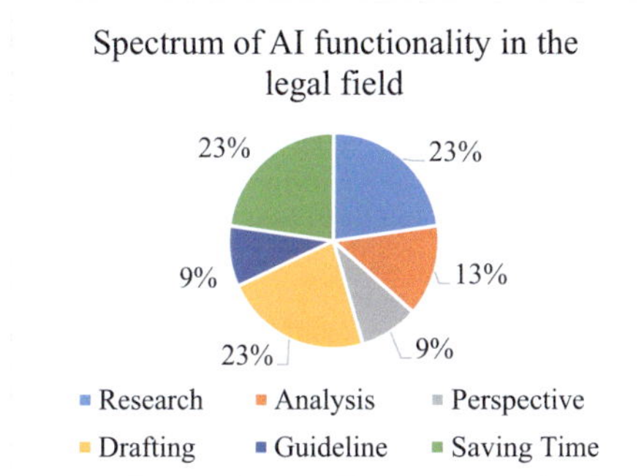

Fig. 2 The spectrum of AI functionality in the legal field

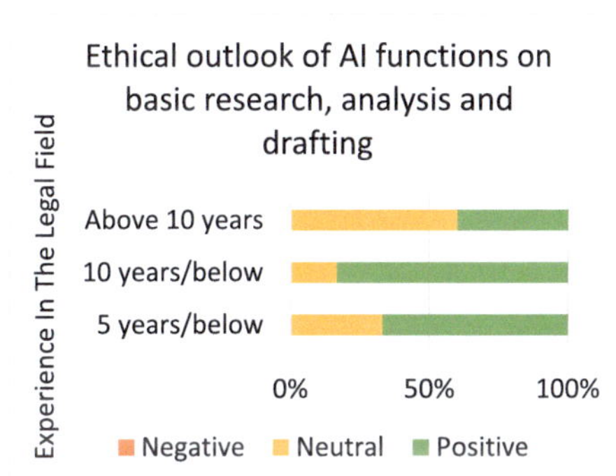

Fig. 3 Ethical outlook on AI functions on basic research, analysis and drafting in the legal field

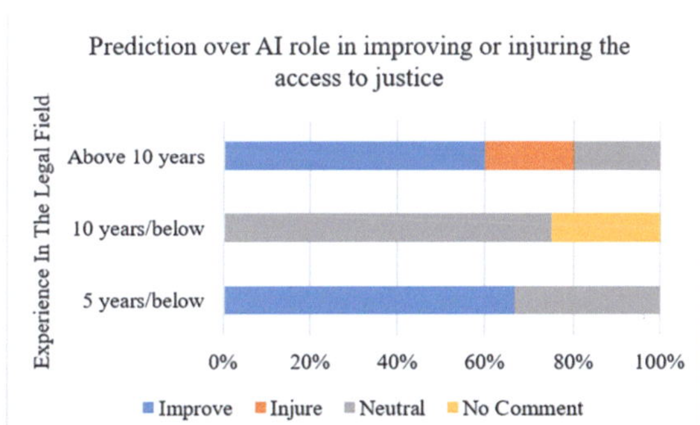

Fig. 4 Prediction over AI role in improving or injuring the access to justice

Figure 5 is complimentary to Figs. 3 and 4. Although the ethical view of AI usage in the legal practice is positive and has been used for among others, research and drafting, yet the services provided still come with a few downsides. Most respondents agreed that the most common downside is how the AI tools could give no humane depth when providing an analysis. Moreover, due to its nature of receiving any input from other users, its generated answer is susceptible to a bias in the algorithm, and the data collected itself could be in breach of data privacy. Besides that, another common issue is on how to key-in input to get the desired answer. This lack of control over the automated process also contributes to the extraction of the cold analysis.

Figure 6 shows an interesting view on the state of existing legislation governing AI. While many respondents who have been in the legal field for more than 10 years expressed their hesitant acceptance to AI usage in the practice, their opinion on the currents laws on AI is the most diverse, with some have the common believe that the laws are insufficient, others believe the law is sufficient, unnecessary or that they have no idea about it.

These findings align with existing research that underscores the potential and pitfalls of AI in legal contexts. The concerns about ethical implications and the call for clearer regulations echo the sentiments found in the literature, advocating for a more structured approach to integrating AI in legal practices.

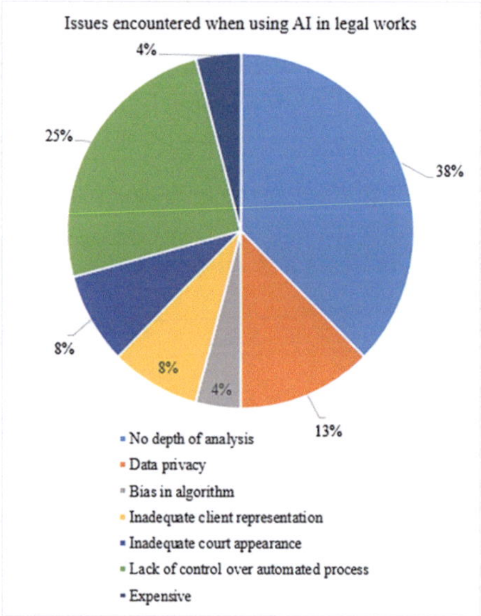

Fig. 5 Issues encountered when using AI in the legal works

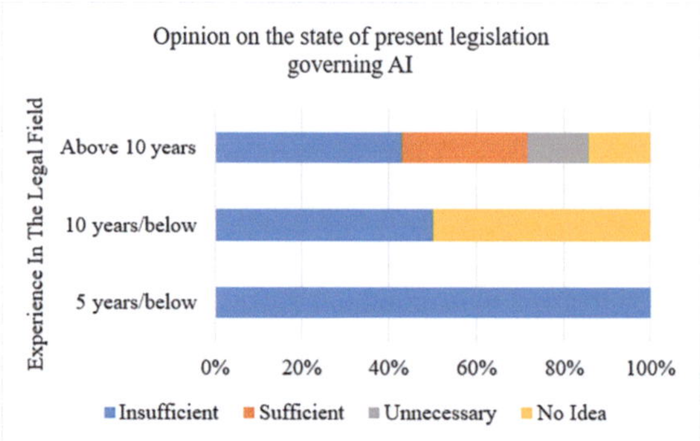

Fig. 6 Opinion on the state of present legislation governing AI

The results suggest that while AI can significantly enhance the efficiency of legal practices, there is a crucial need for frameworks that address the ethical and procedural challenges it presents. Legal institutions may need to consider developing training programs and guidelines to better prepare legal professionals for the technological shifts in their field.

4 Conclusion

Based on the answers received from the respondents, it is obvious that legal practitioners do not use a wide range of AI tools except from ChatGPT and few others. The use of AI is mainly focused on gathering information and to form simple, straightforward links between the information gathered. These newly assembled and organized set of information is then used by the respondents, either as a guideline or draft, for them to further elaborate. Hence, although some hold on to their reservations, most respondents have a neutral to positive view on the notion of AI use in the legal practice. This study has shown that AI is changing the legal profession. Lawyers are using AI more and more to help with their work, which can make things faster and save cost. However, using AI is not as straightforward as it might seem. It brings up a lot of big questions, especially about fairness and whether everyone has equal access to justice. Our survey found that while AI can do a lot of the routine work in law firms, there are worries about whether it can handle more complex legal thinking. Lawyers are also concerned about how ethical it is to use AI, and they are not sure if AI might end up making some of the existing problems in law worse, like biases in decision-making.

References

1. G. Mercan, Z. V. Selcuk, Artificial intelligence (AI) activities in legal practices. Int. J. Eurasian Educ. Cult. **9**(25), 131–144 (2024). http://dx.doi.org/ https://doi.org/10.35826/ijoecc.1824
2. N. Gaubienė, Can artificial intelligence engage in the practice of law as the art of good and justice? Filos. Sociol. **35**(2), 54–63 (2024). https://doi.org/10.6001/fil-soc.2024.35.2Priedas.Special-Issue.6
3. A. Fabregat-Hernández, J. Palanca, V. Botti, Exploring explainable AI: category theory insights into machine learning algorithms. Mach. Learn. Sci. Technol. **4**(4) (2023). https://doi.org/10.1088/2632-2153/ad1534
4. G. Said, K. Azamat, S. Ravshan, A. Bokhadir, Adapting legal systems to the development of artificial intelligence: solving the global problem of AI in judicial processes. Int. J. Cyber Law. **4** (2023)
5. A. Završnik, Algorithmic justice: algorithms and big data in criminal justice settings. Eur. J. Criminol. **18**(5), 623–642 (2021). https://doi.org/10.1177/1477370819876762
6. J. Dressel, H. Farid, The accuracy, fairness, and limits of predicting recidivism. Sci. Adv. **4**(1), 1–5 (2018). https://doi.org/10.1126/sciadv.aao5580
7. G.H. Kasap, Can artificial intelligence ('AI') replace human arbitrators? Technological concerns and legal implications. J. Disput. Resolut. **5**(2), 209–253 (2021)

8. Y. Razmetaeva, N. Satokhina, AI-based decisions and disappearance of law. Masaryk Univ. J. Law Technol. **16**(2), 241–267 (2022). https://doi.org/10.5817/MUJLT2022-2-5
9. A. Zafar, Balancing the scale: navigating ethical and practical challenges of artificial intelligence (AI) integration in legal practices. Discov. Artif. Intell. **4**(1) (2024). https://doi.org/10.1007/s44163-024-00121-8
10. C. Rocha, J. Carvalho, Artificial intelligence in the judiciary: uses and threats, in *CEUR Workshop Proceedings*, p. 3399 (2022)
11. M. Daud, Artificial intelligence in the Malaysian legal system: issues. Chall. Way Forw. **39**(1), 1–24 (2022)
12. N. Grgic-Hlaca, E. M. Redmiles, K. P. Gummadi, A. Weller, Human perceptions of fairness in algorithmic decision making: a case study of criminal risk prediction, in *Web Conference 2018—Proceedings of the 2018 world wide web conference 2018*, pp. 903–912 (2018). https://doi.org/10.1145/3178876.3186138.
13. R. Matulionyte, A. Hanif, A call for more explainable AI in law enforcement, in *Proceedings EEE 25th International Enterprise Distributed Object Computing Workshop (EDOCW)*, pp. 75–80 (2021). https://doi.org/10.1109/EDOCW52865.2021.00035
14. Y. Qu, Z. Zhang, B. Bai, The way forward for legal knowledge engineers in the big data era with the impact of AI technology, in *2023 6th International Conference on Artificial Intelligence and Big Data, ICAIBD 2023*, pp. 225–229 (2023). https://doi.org/10.1109/ICAIBD57115.2023.10206169
15. J. A. Jalil, S. M. Sherif, Legal tech in legal service: challenging the traditional legal landscape in Malaysia. IIUM Law J. **28**(1), 279–301 (2020)
16. J. O. McGinnis, R. G. Pearce, The great disruption: How machine intelligence will transform the role of lawyers in the delivery of legal services. Fordham Law Rev. **82**(6) 3041–3066 (2014). https://doi.org/10.21202/1993-047x.13.2019.2.1230-1250
17. A. E. Davis, The future of law firms (and lawyers) in the age of artificial intelligence. Rev. Direito GV. **16**(1) 1DUMMT (2020) https://doi.org/10.1590/2317-6172201945
18. J. B. Rajendra, A. S. Thuraisingam, Artificial intelligence and its impact on the legal fraternity. UUM J. Leg. Stud. **13**(2), 129–161 (2022). https:// doi.org/https://doi.org/10.32890/uumjls2022.13.2.6
19. T. Hu, H. Lu, *Study on the Influence of Artificial Intelligence on Legal Profession*, vol. 110, pp. 964–968 (2020). https://doi.org/10.2991/aebmr.k.191225.184

MIX
Papier aus verantwortungsvollen Quellen
Paper from responsible sources
FSC® C105338

FSC
www.fsc.org

If you have any concerns about our products,
you can contact us on
ProductSafety@springernature.com

In case Publisher is established outside the EU,
the EU authorized representative is:
Springer Nature Customer Service Center GmbH
Europaplatz 3, 69115 Heidelberg, Germany

Printed by Libri Plureos GmbH
in Hamburg, Germany